Design Criteria for Driving, Pilotage, and Remote Control Imagers

Dr. Richard H. Vollmerhausen

Part 1

This work is taken from public domain, and not copyrighted to author. This sole purpose is to gather all the information at one place.

Part 2

Dr. Richard H. Vollmerhausen

ISBN

Preface

Advancing technology makes piloting and driving under adverse visibility conditions possible. This book provides design guidelines for navigation imagers and explains the basis for those guidelines.

The technical level of this book is oriented toward a reader with an engineering background, but expertise in imager design is not needed, and the book's mathematical level is limited to basic algebra. The book's emphasis is on function and not theory. However, the book provides imager design details and the physical parameters needed to specify or model imaging hardware. This book is not intended for a general audience.

The material in Part 1 of this book is in the public domain. The work described in Part 1 was done by government employees and published in the collection of papers, reports, book chapters, and articles listed in the bibliography. This book re-visits the published material in order to make it readily available and also to provide a holistic view of the combined data.

Part 2 of this book describes a value-added package to add color and thermal to a PVS-14 goggle. The concept provides high resolution image intensifier, color, and thermal imagery while adding minimal weight and cost to the PVS-14. The upgrade has application for military, police, hunting, search and rescue, any application that requires improved night vision capability.

Table of Contents

Part 1

Design Criteria for Driving. Pilotage, and Remote Vehicle Control Imagers

Chapter 1
Introduction

This book is about designing navigation imagers. A navigation imager might be used by a driver or pilot to visually drive or fly at night, or the navigation imager might be a camera on an unmanned vehicle with the operator located remotely. A well-designed navigation imager lets the operator control the vehicle "head's up and eyes' out" in a manner that approaches day-like vehicle control at night, under poor weather, or from a remote location.

We will explore existing night vision systems, user feedback on those systems, as well as the minimum illumination and thermal scene contrasts under which systems must operate to garner user acceptance. That information is useful when designing advanced systems, because fielded systems probably establish a minimum performance standard.

This book also provides a wealth of technical data. Flight experiments to establish pilotage design criteria are described and the results discussed. Parameters that quantify the performance of near infrared and short-wave infrared solid state photo detector arrays are compared to image intensifier characteristics, and the potential of replacing image intensifiers with a staring, solid state imager is discussed. The benefits of the long wave thermal spectral band versus mid wave thermal are described and quantified.

Four subjects receive special attention. The first is motion blur. Fielded systems that use outdated technology have negligible motion blur. However, modern solid state staring imagers will exhibit excessive motion blur in the navigation application unless steps are taken to mitigate the problem. This book describes the origins of the motion blur.

There appears to be a widespread misunderstanding that mid and long wave thermal imagers are equivalent for terrestrial applications. That is not the case. The origins of the misunderstanding are described, and the benefits of long wave are quantified.

Part 1 describes a way to replace the helmet display with a panel display while retaining full helmet display functionality. Eliminating the cost and bother of a fitted helmet, helmet display, and hours of training will broaden the application of navigation imagers.

The forth subject that receives special attention is a method of fusing low resolution, high sensitivity color and thermal images with a high resolution goggle image to achieve high resolution color and thermal night imagery, Since the color and thermal cameras have a low pixel count, the sensitivity is high, but the cost and weight are low. The color and thermal goggle is described in Part 2 of this book.

Description of Navigation Imagers

The navigation imager design criteria apply to vehicle control concepts like those illustrated in Figures 1.1 through 1.5.

In Figure 1.1, the pilot is in the rear seat of the aircraft but the imager is in a turret on the nose. A turret is a pan tilt mechanism that can point the imager at any angle within the angular limits indicated in the figure. A helmet tracker points the imager where the pilot wants to look, and video from the imager is relayed back to a display mounted on the pilot's helmet.

The pilot sees a unity magnification scene in his right eye. Unity magnification means that objects appear life-sized; that is, objects are the same size as when viewed with the naked eye. Unity magnification correctly presents visual motion and range cues.

Figure 1.2 is another example of a navigation imager. The portion of the scene imaged by the camera is called the field of view (FOV), and the FOV is indicated by the green box. The turret can point the camera over a much larger portion of the scene surrounding the vehicle, and that larger scene area is called the field of regard (FOR). The turret is coupled to the head motion of the operator, so imagery is collected from the direction in that she points her head. The video from the turret imager is fed back and viewed on a helmet display.

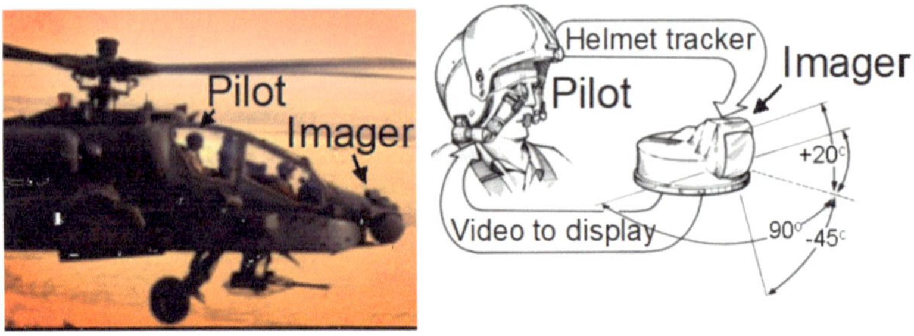

Figure 1.1. At left, the pilot is in the rear seat, and the imager is on the nose. As illustrated at right, the imager is in a turret that is directed in space by the pilot's head. The video from the nose is fed back to a display on the pilot's helmet. He sees the image only in his right eye.

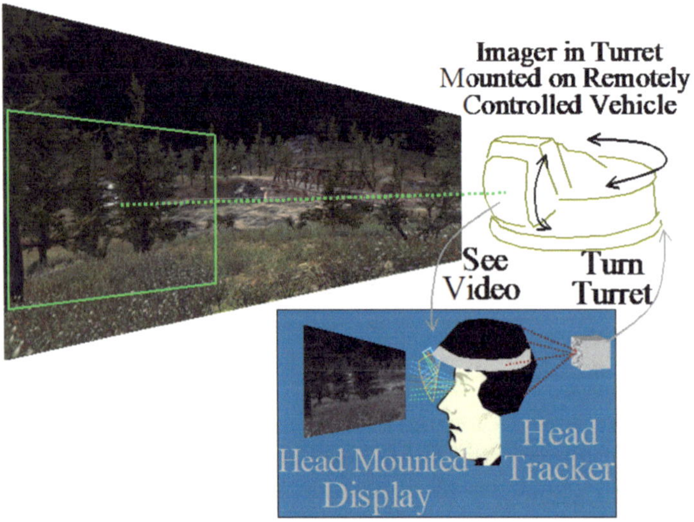

Figure 1.2. Like the Figure 1.1 configuration, the imager is pointed based on the observer's head position, and the imager's video is coupled back to the observer.

In Figure 1.3, the camera has a very wide FOV that cannot be displayed in its entirety on the helmet. The camera FOV is now the operator's FOR. The helmet tracker is used to select a portion of the camera FOV. The benefit of this approach is that the turret is not needed, and the FOV follows the operator's head almost instantaneously.

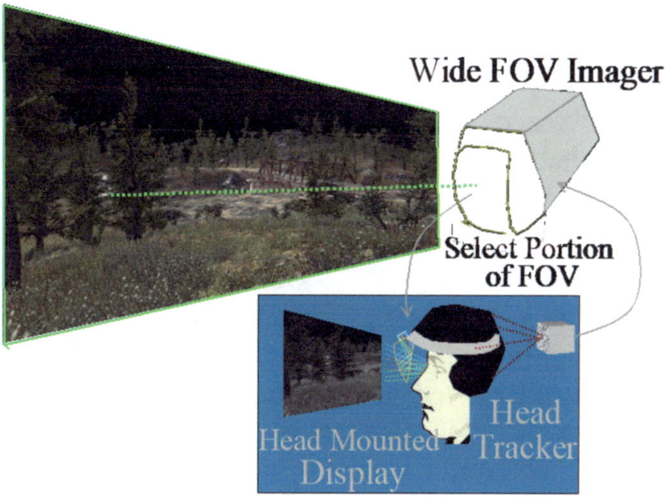

Figure 1.3. The turret is not needed if the camera FOV is very large. The camera FOV is indicated by the green box surrounding the scene. The camera FOV becomes the system FOR, because the operator can only see the smaller FOV provided by the helmet display. The head tracker is used in the same way as in Figure 1.1 and 1.2, except the turret lag is eliminated, and multiple aviators/drivers/operators can use the same camera.

In order to render the imagery correctly, the helmet display must be high quality, and it will be a significant fraction of the navigation system cost. Also, keeping the display optics aligned with the eyes' exit pupil is necessary and requires a fitted helmet. Further, using the helmet display for hours can lead to eye strain, and the helmet display is the primary limiter of implementing a wide FOV.

The helmet display is probably the most unattractive feature of the navigation systems shown in Figure 1.1 through 1.3. However, the helmet display provides some critical features that are not provided by a fixed forward camera viewed on a panel display.

A fixed forward camera viewed on a panel display does not substitute for a "window." For one thing, scene magnification changes markedly as the observer moves his head closer to the display, whereas scene magnification would change little when looking through an actual window. Also, the FOV gets bigger when the head gets closer to a window, whereas the image on a display remains fixed.

Another difference between a fixed forward camera and a window is that moving the head side-to-side when viewing through a window lets the observer see to either side of the vehicle's path, whereas the image on a fixed forward camera does not change.

The configurations illustrated in Figures 1.4 and 1.5 change the display into a virtual window. Many forms and sizes of flat panel displays are available today, including displays with controlled transparency and displays that can be curved. The display can be molded to the window or windows, or one or more displays can be dropped down or folded out in front of the pilot/driver when needed.

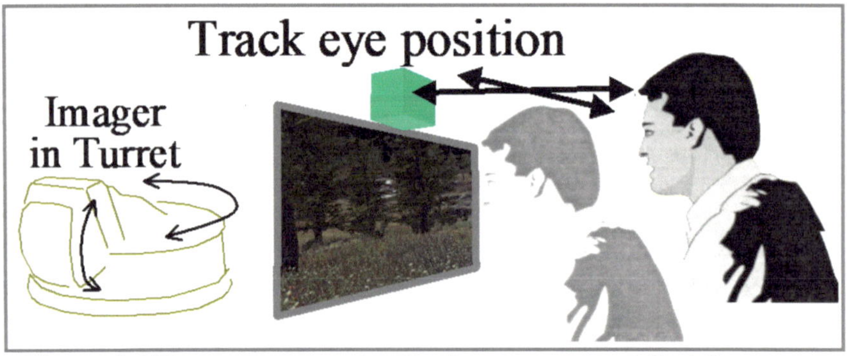

Figure 1.4 illustrating a navigation imager concept that does not require a helmet display.

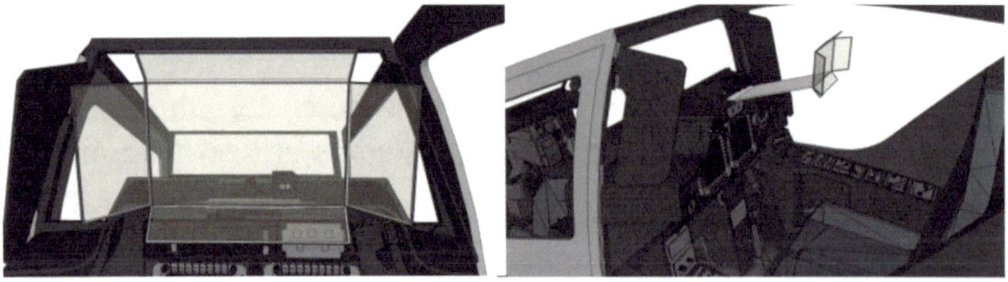

Figure 1.5 shows two options for display size and positioning.

Both imager magnification and turret pointing can be controlled by tracking the position and alignment of the two eyes. (That is, the location of eye sockets or glasses relative to tracker, and not movement of eyes within the head.) See Figure 1.6.

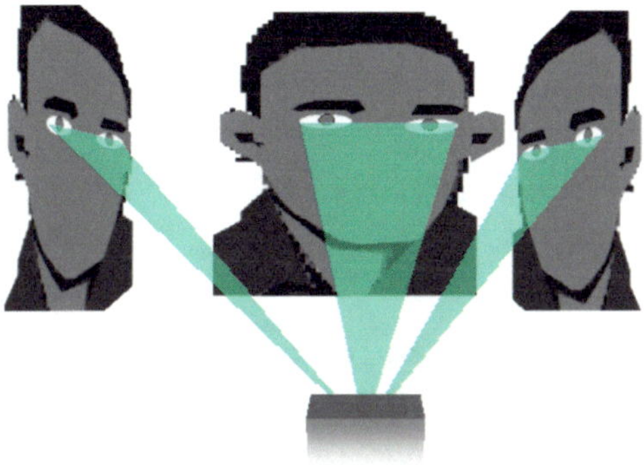

Figure 1.6 showing device tracking the location of eye sockets.

Scene magnification is changed electronically based on the eye to display distance. Turret pointing is based upon three parameters obtained from the eye tracking data, namely the eye to display distance, left-right position of the head, and the change of apparent eye separation and angle due to head rotation. All of the needed geometry information can be obtained by tracking the position of the two eye sockets (or eye glasses or sun glasses) relative to the display.

The navigation imager concept illustrated in Figures 1.4 through 1.6 does not use a helmet display but still provides unity magnification of the scene plus a large FOR. Also, when using this navigation imager configuration, it is much easier to expand the FOV. Ergonomics factors make it difficult to expand a helmet display FOV, whereas with a panel display, FOV is increased by using a larger monitor or by placing the viewpoint closer.

This book focuses on the imaging sensor and the effect of imager characteristics on vehicle control. That limited scope reflects the author's expertise and is not meant to disparage the importance of symbology.

Also, the material in Part 1 of this book does not apply to fixed forward thermal imagers displayed with minified, non-unity, magnification. There are too many functional differences between fixed forward and head tracked system to treat them together. The Part 1 material applies only to systems of the kind illustrated by Figures 1.1 through 1.6.

Image intensified goggles are a special case where the imager is mounted on the user's helmet. Night vision goggles are described in Chapter 2.

Overview of Navigation Imager Design Requirements

While it is true than we can run into things we do not see, like wires, it is also true that we can run into things that we do see, like sand dunes, if the imager provides inaccurate or inadequate motion cues. The ability to see a hill, and the ability to correctly judge that the current flight path clears the hill, are two different visual tasks.

Simply providing a visual cue that something is in front of the vehicle is not the function of a navigation imager. The imagery should let the operator comprehend terrain slope, sense vehicle motion, and quickly and intuitively avoid closure on surrounding objects. The visual cues used by the vehicle operator are stimulated by motion parallax, and motion parallax is generated by the texture of local objects.

Flying over sand dunes in Saudi Arabia is not the same visual task as flying over rolling hills in the eastern United States. A sand dune can be equated to a sheet of white paper, whereas hills in the eastern U.S. have grass, bushes, trees, and rocks, things that create texture to stimulate motion parallax. Of course, the navigation imager must have sufficient resolution to render the cues provided by the terrain.

On your first flight using night vision goggles, the instructor pilot takes the aircraft a few feet off the ground and flies slowly toward what appears to be six inch high ground cover. Ten or twenty feet from the ground cover, it pops up and becomes three foot high bushes. You jump, surprised, and the instructor laughs. Happens to everyone.

What just happened? On that bright, clear night, the goggles were providing their best resolution. Still, when the aircraft was far from the bushes, the blurred picture presented to the eye caused you to misinterpret the nature of the terrain in front of the aircraft. The bushes looked like flat ground cover. When the aircraft got close enough for your brain to correctly interpret the visual cues presented by the goggles, what appeared to be six inch high ground cover suddenly became three foot high bushes.

Camera resolution depends, of course, on the quality of the lens and on the camera pixel count. However, for navigation imagers, a third factor is also crucial. That is, camera exposure time, sometimes called dwell time. Exposure or dwell time is the period spent collecting light on the photo detectors. Short exposure times are used to capture pictures of sports events, because there is rapid motion in the scene. Navigation imagers also need to capture pictures of rapidly moving scenes.

Consider pictures A, B, and C in Figure 1.7. A and B are taken from a helicopter at Nap of the Earth (NOE) height, and picture C is taken from a helicopter flying contour. NOE is flight below the height of nearby obstacles, whereas during contour flight, the aircraft is maintained at constant height above local obstacles.

Figure 1.7. Pictures A and B were taken from a helicopter flying Nap of the Earth, and the aircraft was below the height of surrounding obstacles. Picture C was taken during contour flight, where the pilot keeps the altitude a fixed distance above the trees.

In Picture A, simply showing that there are trees past a small hill in the flight path is not good enough. The aviator wants to maneuver the aircraft through the surroundings, going over the small hill and past the trees. But that means seeing the slope of the hill and the distance and closure on the trees. In Picture B, the parallax motion cues are provided by the tree limbs. In NOE flight, the aircraft is traveling slow, probably no more than thirty knots. However, terrestrial objects are close by, and the scene seen by the aviator changes rapidly.

In C, the objects of interest are the impending trees. The helicopter is heading toward a tree line at 140 miles per hour, a tree line that the aviator hopes will end up 30 feet below the aircraft. The function of the navigation imager is to help the aviator judge his closure on the impending trees. The pilot is flying low and fast on purpose, in order to avoid anti-aircraft fire, and he knows the trees are there. What he needs from a night vision aid is the ability to judge distance and closure on the nearby and fast approaching trees.

The boughs and branches of the trees in Figure 1.7 C form a complex three-dimensional surface. Parallax motion cues are generated by the rapidly changing perspective of that surface as the aircraft proceeds along its flight path. The role of the navigation imager is to render those rapidly changing motion cues.

The navigation imager should help the pilot to quickly find a safe flight path while executing contour and NOE flight. There is a reason that the aviator is not flying at high altitude, but he will climb if flight safety requires it. That judgment, what altitude is the safest, depends on many factors, and one of those factors is the quality of the navigation imagery, and the ability of the navigation imager to render scene movement.

The navigation imager must provide a wide field of view while rendering scene texture (the bushes and tree branches) while tracking the operator's head while he is driving or flying a moving vehicle past nearby objects. We know from the analysis presented in Chapter 4 that long photo detector exposure times are not compatible with the movement of a car or helicopter through close surroundings. We know from experience that operation in close surroundings is a common occurrence. However, modern, solid state, night vision imagers typically achieve near theoretical performance by using long exposure times.

We are not suggesting that solid state, staring array solutions are not suitable for navigation imagers. We are suggesting that adapting that technology to the navigation imager application will require a combination of clever engineering and compromise.

What is in Part 1 of this book

Chapter 2 describes existing head-tracked pilotage and driving aids. Both systems were fielded in the early 1980's, and advanced versions of both systems have been in the field for over twenty years. Those systems provide design examples of successful navigation imagers. Also, those systems provide the standard upon which replacement systems will be judged.

Part 1of this book provides design guidelines for navigation imagers and explains the logic supporting those guidelines. The design guidelines might be based on physical data or flight experiments, but often the navigation design criteria are based on user experience. It is therefore appropriate that Chapter 2 describe the currently fielded navigation imagers.

Chapter 3 provides background on the staring imager technology that is expected to replace existing systems. Generic staring imagers are described, and the parameters used to quantify imager operation are defined. Fourier analysis concepts and utility are explained in Chapter 3 without using calculus. Chapter 3 also describes the radiometric units used in subsequent chapters.

Chapter 4 presents some analysis of the effect of motion blur on navigation imagery. That chapter also describes the results of flight evaluations that were conducted to learn how motion blur impacted helicopter flight control. Motion blur is a critical consideration for navigation imagers, but the analysis details could not be presented before describing Fourier concepts in Chapter 3.

Chapter 5 discusses factors that distinguish thermal and reflective imager designs. The difference between long wave infrared and mid wave infrared scene contrast is described, and the benefits of long wave thermal for navigation imagery are quantified. Some of the benefits of mechanically scanned imaging are discussed.

Chapter 6 provides data on night illumination in the visible, near infrared and short-wave infrared spectral bands. The problems with replacing image intensifiers with either near infrared or short-wave infrared staring arrays are discussed.

Beginning with **Chapter 7**, each chapter focuses on a different design parameter. Chapter 7 describes FOV versus resolution trades. Helmet mounted goggles were fabricated with various combinations of FOV and resolution; see Figure 1.8. The subject aviators flew front seat Cobra in order to minimize the effect of aircraft obstructions on experimental results. The subjects were senior aviators, all recognized by their peers for their flight skills and experience. Realistic flight profiles, established by the aviators, tested alternative imager design options. The Cooper-Harper evaluation methodology was modified such that all questions were comparisons between flight configurations.

Figure 1.8 shows direct view goggles fabricated to test various combinations of FOV and image resolution. The aviator subjects flew from the front seat of a Cobra.

Chapter 8 describes flight experiments that used eye and head tracking to explore what happens when each eye sees a different part of the FOV. At far left in Figure 1.9, both eyes see the same image. At center in the figure, the left eye (LE) and right eye (RE) both see the same central part of the FOV. However, the two eyes see different parts of the scene at the two peripherals of the FOV. The picture at right shows an aviator with the eye tracking hardware in place. The aviator's eyes were masked to provide the desired FOV with overlap, and a flat plate eye tracker was modified to operate in sunlight.

Chapter 8 explains the cause of ‘looning” at the edge of the central, binocularly viewed region of the FOV. Chapter 6 also presents the eye tracking data that shows that the eye is essentially “trapped” within the central, biocular region of the FOV.

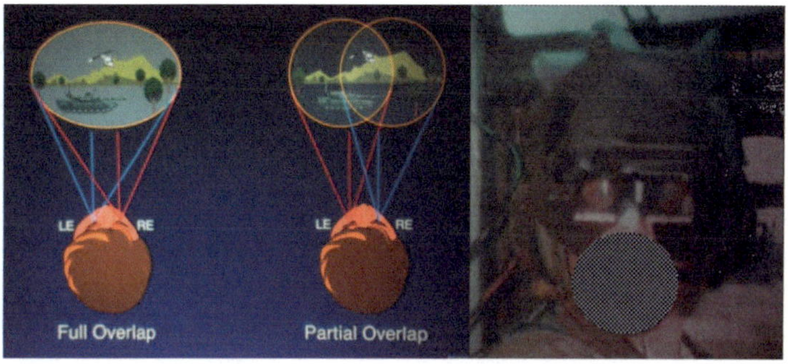

Figure 1.9. The flight experiment used eye tracking and head tracking to explore what happens when each eye sees a different part of the FOV.

Chapter 9 describes three flight evaluations to show users the effect of questionable design compromises. The first flight demonstrated what happens when the imager video is delayed in digital processors before being displayed to the operator. In modern aircraft, the imager, data link interface, avionics computer, symbology generator, and display electronics are all provided by different vendors. For convenience, each video handoff between vendor might call for a video field or frame delay, and contracting convenience leads to serious control problems.

The flight evaluation used the head mounted camera shown in Figure 1.10. Once again, the Cobra Helicopter was used in order to minimize the impact of aircraft structure. A cloth over the pilot’s helmet prevented direct view from the aircraft. Video presentation on the helmet display was delayed using a digital frame grabber. The aviators performed a series of realistic flight tasks, and answered the same Cooper-Harper-like questions used for the FOV versus resolution experiment.

The second flight demonstration used the hardware shown in Figure 1.10 with different camera and digitizer settings to evaluate the effect of display dwell time on flight control. The program needed to specify a maximum pixel dwell time for a new helmet display that employed a flat panel display rather than a CRT.

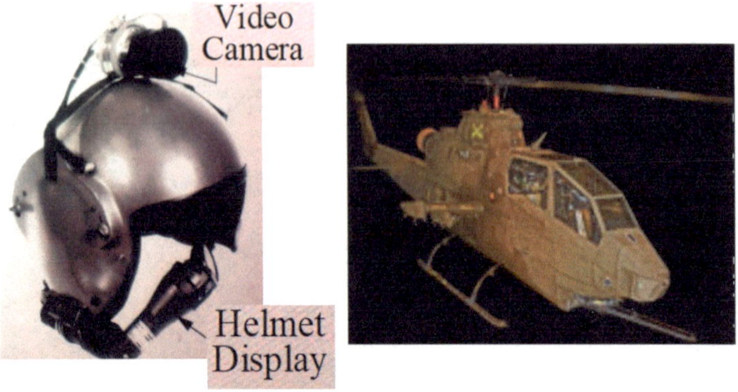

Figure 1.10 showing the helmet mounted camera used to capture video.

The third topic in Chapter 9 describes flight demonstration of imagers using field replication. When using field replication, the camera operates at 30 Hertz progressive, but display at 30 Hertz would result in flicker. Therefore, the 30 Hertz camera video is displayed twice, $1/60^{th}$ of a second apart. Field replication leads to seeing double when the camera is panned or something in the scene moves. The double images are quite apparent and visually disruptive when viewed at unity magnification. Multiple demonstrations used the hardware shown in Figure 1.10 and also the thermal imager, helmet display, and head tracking shown in Figure 1.11.

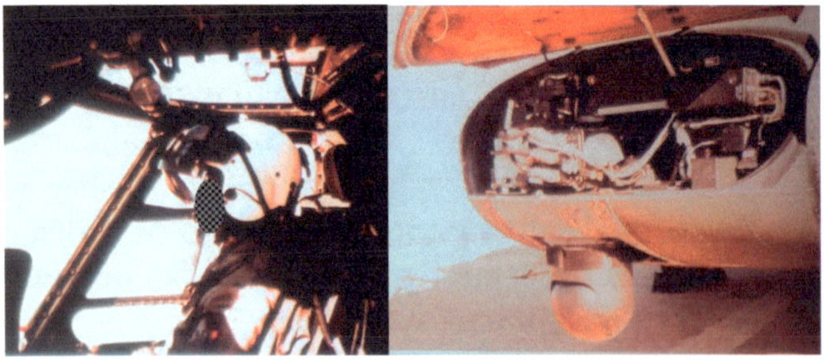

Figure 1.11 showing one of the flight systems used to demonstrate the effect of video field replication.

Chapter 10 evaluated the aviator's ability to resolve imagery on a panel display. Sometimes a pilot flying with ANVIS uses a non-head-tracked, hand-controlled thermal imager to double check his flight path. See the left picture in Figure 1.12. The question arose as to whether helicopter vibration would prevent the aviator from seeing the details provided by a high resolution, thermal imager.

The aircraft's rotors were unbalanced to the "call for maintenance" point, and the aircraft instrumented with accelerometers. Then the back lit, multi-contrast bar charts shown at right in the figure were used to test the aviator's resolution of the panel display.

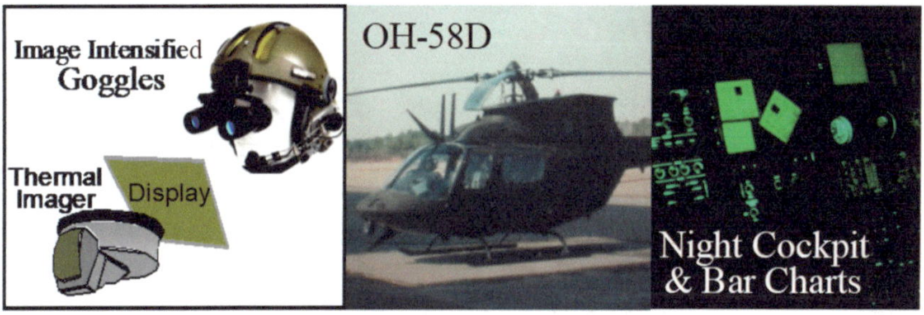

Figure 1.12 shows the set up to test aviator's ability to see panel display details, at night, when wearing ANVIS.

What is in Part 2 of this book

Part2 describes a way of adding both color and thermal imagery to an image intensifier. The technique described in Part 2 uses an RHV Electro-Optics, LLC patent to implement high resolution color and thermal while adding surprisingly little cost and weight.

Adding color and thermal to the image intensifier imagery greatly improves situation awareness. The goggle will find application to the military, of course, but also police, security patrols, hunters, search and rescue, and other applications that use reasonably priced night vision aids.

We are searching for a company with the engineering skills and resources to execute what should be a short development cycle. We are on the web at RHVElectro-Optics.com.

Chapter 2
Fielded Navigation Systems

The currently fielded systems have been operational for over twenty years, and both systems receive adequate to good ratings from the user community. Both systems provide design examples of successful navigation imagers, and these systems likely establish minimum performance standards for future replacements.

Image Intensifiers

The most common pilotage and driving aids are image intensifiers as represented by the Aviator's Night Vision Imaging System (ANVIS). ANVIS is shown at right in Figure 2.1; it is mounted on a helmet. ANVIS is a binocular system with a separate image intensifier for each eye, but many of the image intensifiers in use are monocular. Image intensifiers amplify ambient light, that is, moonlight and starlight, and they brighten the dark night scene so it is visible to the eye.

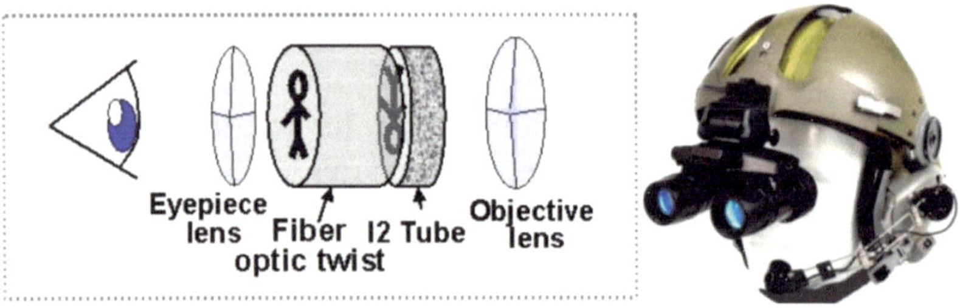

Figure 2.1. ANVIS at right and components in each ocular at left.

The operator is inside a vehicle, and light from panel instruments would normally saturate the image intensifier. However, the 3rd Generation cathode material responds mainly in the near infrared, and color panel displays of course emit visible light. Spectral filters on each ANVIS ocular block visible light emitted by the aircraft's instruments. On a dark night, pilots will set instrument panel luminance in the vicinity of a foot Lambert, and that mesopic light level allows the pilot to see the outside scene through ANVIS, but he can look under ANVIS and have color vision of the instruments with nearly photopic visual acuity.

A schematic of one direct view goggle ocular is shown at left in Figure 2.1. The objective lens forms an inverted image of the scene on the image intensifier tube. The image intensifier tube amplifies the brightness of the image as described below. The fiber-optic twist erects the brighter image. The eyepiece creates a unity magnification, virtual image of the scene, allowing the pilot to fly at night without lights.

Operation of the image intensifier tube is illustrated in Figure 2.2. Photons from the scene generate photo-electrons in the cathode. A high voltage accelerates the photo-electrons to the micro channel plate (MCP). The MCP consists of millions of tiny channels; these channels are about five microns in diameter on a pitch of six microns. The channel length to diameter ratio is about seventy.

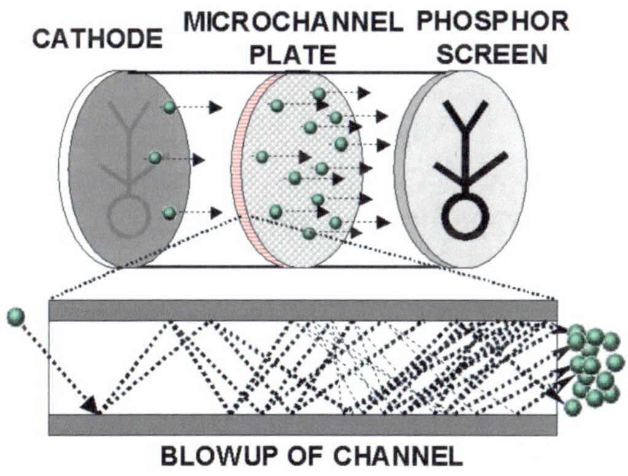

Figure 2.2. Illustration of image intensifier tube operation

Operation of the MCP is shown by the blowup of a single channel at the bottom of the figure. Photo-electrons enter the channel and are accelerated by a high voltage across the channel plate. Secondary electrons are emitted when the photo-electrons strike the channel wall. The secondary electrons are then accelerated, strike the wall, and create more electrons. Electron gain through the channel is controlled by varying the voltage across the MCP. Channel electrons exit the MCP and are accelerated by another high voltage to the phosphor where an image is formed.

Brightness gain results from the MCP electron gain, the energy gained from electron acceleration between the MCP and phosphor, and from the fact that the cathode is sensitive to a much broader range of light wavelengths than the eye. Brightness gain is specified by the ratio foot Lamberts from the phosphor to foot candles on the cathode. Typical gain is 30,000 but gains to 200,000 are possible; however, gains in excess of 100,000 leads to bothersome scintillation in the image. Brightness output of the image intensifier tube is controlled by limiting the current available to the MCP; generally, goggle brightness is limited to about 3 foot Lamberts.

Early versions of ANVIS also controlled gain by lowering the cathode to MCP voltage, but that led to image blur. Modern ANVIS gates the cathode to MCP voltage on and off, so the image intensifier provides a good image under sunlight or starlight. However, continual operation in sunlight would likely hurt the cathode.

Early versions of ANVIS had halos around bright objects, because some cathode electrons would bounce off the MCP, back to the cathode, and then bounce back to the MCP. That problem has also been corrected.

Tube noise factor is ideally 1.4 based on the open area ratio of the MCP. (The open area ratio is the area of holes divided by total MCP area. Not all photo-electrons from the cathode can get through the MCP.) However, the typical noise factor is about two, meaning that the image intensifier noise is a factor of two worse than ideal.

Image intensifiers make the scene brighter with virtually no signal storage, and that means no motion blur. Noise due to electron amplification is only a factor of two worse than what is theoretically possible. The ratio of foot Candles on the image intensifier cathode to foot Lamberts phosphor luminance can be 100,000 without exhibiting excess noise. That light gain lets us see at night, even in some overcast starlight, well enough to drive a car or fly a helicopter. Further, modern ANVIS with gated power supplies provide artifact-free, useful imagery with scene illuminations that can vary more then five orders of magnitude.

Head-Tracked Thermal Imager

The AN/AAQ-11 Pilot's Night Vision System (PNVS) was developed for the AH-64 Apache Advanced Attack Helicopter. Both the 1st Generation and 2nd Generation systems will be described so that the reader can correlate user experience with an imager design.

Most of the details currently in the public domain describe the 1st Generation system. The 1st and 2nd Generation imager designs are very different, but most of those details reflect the evolution of electronics technology from the mid 1970's to the mid 1990's. Features like automatic line-to-line uniformity adjustment, for example, are important in fielded systems, and no imager designed today would lack that feature. Since digital image processing was not available when the 1st Generation system was designed, that imager lacked many of the features that would be considered standard today.

This section focuses on three design topics. First, the 2nd Generation system provided a much needed improvement in imager sensitivity and resolution. Second, the advanced system continued to use mechanical scanning to form an image, and that retained the 1st Generation's capability of imaging rapidly moving scenes. Third, the advanced system retained the CRT in the helmet display, and that retained the speed, brightness, and dynamic range capabilities of that older technology.

The PNVS senses thermal energy, that is, heat radiated from the surroundings. The original PNVS imaged 8 to 12 micron thermal energy using a linear array of photo diodes. Table 2.1 lists hardware parameters for the 1st generation PNVS. Table 2.1 also lists the imager parameters for a prototype 2nd Generation system called Advanced Helicopter Pilotage (AHP).

Table 2.1. Thermal Imager Parameters

Parameter	1st Generation PNVS	AHP Prototype
Spectral band	7.5-12 microns	8-11 microns
Field of view	30 V by 40 H	30 V by 40 H
Detector array	180 by 1 HgCdTe PC	480 X4 TDI HgCdTe PV
Detector size	41 by 60 microns	35 by 50 microns
Sampling (H/V)	Analog / 1.6 per dwell	2 / 2 per dwell
D-star	8E10 Jones	1.5E11 Jones
F/number	1.52	1.75
Focal length	3.33 centimeters	2.2 centimeters
Frame Rate	60 frames per second	30 frames per second
Vertical Lines	360 Infrared & 720 Video	960 Infrared
Interlace	Yes	Yes
Display Type	Cathode Ray Tube	Cathode Ray Tube

AHP preceded the 2nd Generation PNVS by almost ten years and was built by a different company; AHP is not the modernized, 2nd Generation PNVS. However, AHP testing confirmed that the time delay and integration technology would fix the sensitivity problem with the original PNVS.

The PNVS was gimbaled and mounted on the nose of the helicopter; the position of the PNVS is shown in Figure 2.3. As illustrated in Figure 2.4, a helmet tracker slaves the sensor's line of sight to the pilot's head. Attached to the helmet is a display through which he views the thermal image. The helmet-mounted display is monocular, viewed with the right eye only, and provides the same 30° vertical by 40° horizontal field of view as the imager. The system provides a unity magnification, thermal image of the surroundings. That is, scene objects appear life-sized.

Figure 2.3. showing the first generation PNVS on the noise of the aircraft.

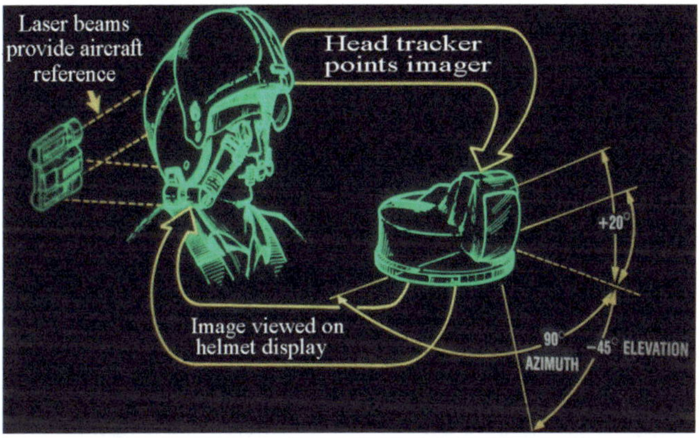

Figure 2.4 illustrating the 1st Generation PNVS navigation system.

The field of regard (FOR) is the solid angle over which the FOV will follow the pilot's head. In Figure 2.4, the turret points plus 20 and minus 45 degrees vertically and plus and minus 90 degrees in azimuth (that is, horizontally). The 30 by 40 degree FOV adds to the FOR, so that the aviator can see 35 degrees up, 60 degrees down, and 110 degrees to either side of the helicopter nose.

The first generation PNVS provided good imagery when the standard deviation of scene contrast was 1 Kelvin. However, scene contrasts less than 1 Kelvin are common. When scene contrast fell below 1 Kelvin, 1st Generation PNVS imagery slowly degraded as contrast fell to about 0.3 Kelvin. Below the 0.3 Kelvin, the imagery rapidly became unusable.

Figure 2.5 shows a histogram of available scene tree-to-ground contrast ratios. The data were taken over several nights in November in Germany. The plot indicates that scene contrast was about 0.1 K for 25% of the time and in the neighborhood or below 0.3 K about half the time. Unfortunately, no description of the weather accompanied the contrast plot.

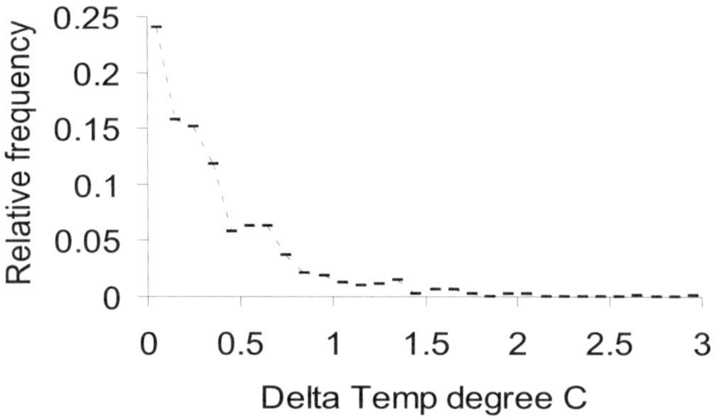

Figure 2.5. Histogram of tree-to-ground thermal contrast taken at night in Germany in November.

We do know from user experience with the 1st Generation system, however, that a couple hours of insolation were needed during the day for PNVS to perform at least adequately throughout the night. The 1st Generation PNVS needed to provide useable imagery under 0.1 Kelvin contrast conditions, and it did not.

Tree-to-ground contrast is measured as follows. A photo detector in the spectral band of interest, in this case 8 to 12 microns, is scanned by a mirror over a large area of the scene. The detector is coupled through a capacitor to an amplifier in order to remove the large, thermal pedestal. The photo detector signal is calibrated using a blackbody. Contrast is defined as the standard deviation of the photo detector signal. Note that it is not necessary for there to be trees in the scene. "Tree-to-ground" is the name of the procedure and not a description of the scene.

Figure 2.6 illustrates the imaging method used by both 1st and 2nd Generation PNVS. A linear array of photo detectors is scanned by a mechanical mirror over the field of view. The 1st generation system used interlace to provide 360 lines of infrared at 60 Hertz frame rate and 120 Hertz field rate. Motion blur was minimized because detector dwell on each point in the scene was quite short.

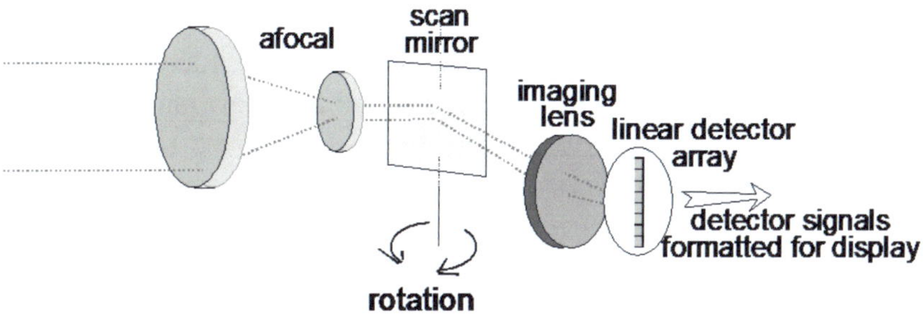

Figure 2.6 illustrating how a linear array of photo detectors is scanned to create a two dimensional image. The 1st generation system used a single linear array of photo detectors.

The 2nd generation system implemented time delay and integration (TDI) to solve the sensitivity problem. See Figure 2.7. TDI uses multiple, linear rows of photo detectors; four rows are shown in the figure. The signal from the first row is stored and delayed so it can be added to the signal from the second row. The combined signals from rows one and two are delayed and added to the row three signal. That process continues until the signals from all four linear arrays are summed, and the summed signal provides the needed sensitivity improvement.

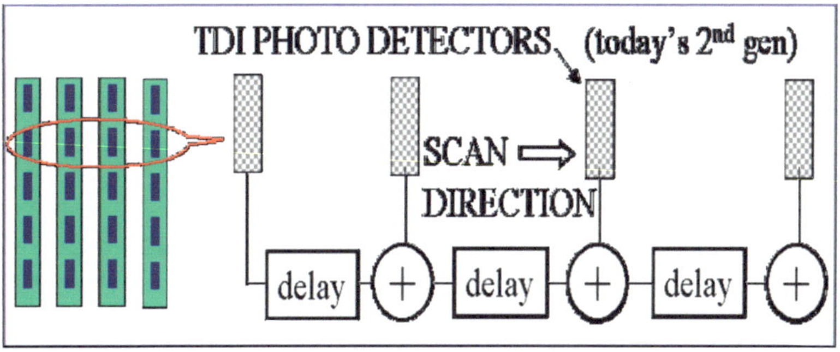

Figure 2.7 showing the operation of time-delay-and-integrate photo detectors.

In the 2nd Generation system, each linear array has 480 photo detectors, and mechanical interlace is used to provide the aviator with 960 line, high resolution video. The 2nd Generation system improved both image resolution and imager sensitivity. Data on the tree-to-ground contrast required by the modernized PNVS is not in the public domain, but published user surveys suggest that the 2nd Generation system solved the sensitivity problem.

Further, an image intensified camera was recently added to the modernized PNVS. Aviators have stated for decades that image intensifiers work better in some areas and thermal imaging works better in other environments. In all cases, the two technologies depend on different physics for imaging and therefore provide different views of the surroundings.

The intensified camera is illustrated in Figure 2.8. A cathode that responds to near infrared radiation emits electrons when illuminated by scene light. The electrons are accelerated to a back-thinned CMOS silicon chip where electron multiplication occurs, and the electrons are collected to form an image.

We do not have information on intensified camera performance.

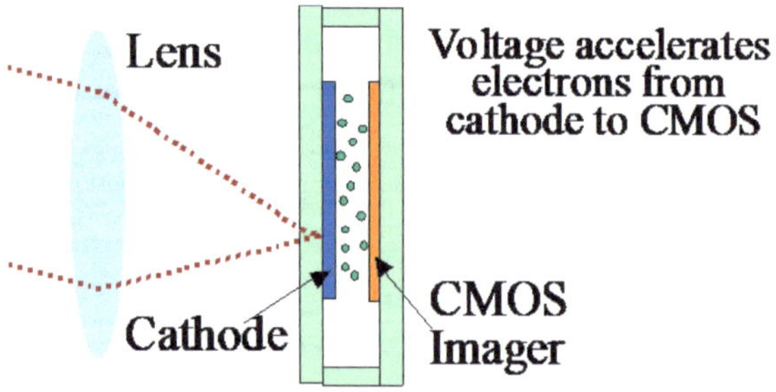

Figure 2.8 illustrates the construction of the modernized PNVS intensified camera.

Some Comments on Replacing Fielded Systems

Over the twenty-five years of field experience with the modernized PNVS, the primary user complaints about the system have been the lack of the recently added image intensifier, eye strain from using the helmet display, and the turret lagging head motion. In other words, although the technology is dated, the thermal imager and CRT helmet display still perform well.

Partly by luck, but mainly by persistent scientific experimentation, solid design decisions, and tight production controls, today's image intensifiers are remarkable devices. Image intensifiers are certainly the most cost effective, best-performing, and easiest to use night vision aid for applications where a directly viewed, head-mounted imager is appropriate.

The two systems described here have provided good service for over two decades, but they are the product of 1980 technologies. The two fielded technologies have been evolved and refined over the years, but imaging technology has moved on.

Although successful in terms of performance, the currently fielded thermal system uses mechanical scanning, and that detector technology is dated and expensive to produce. Certainly a smaller, lighter, and more capable imager can be designed using the wide variety of thermal detector arrays available today.

As discussed in a later chapter, finding a replacement for image intensifiers will be more difficult. Image intensifiers are both sensor and display, they provide excellent imagery, they run on very little power, and because they are head mounted directly in front of the eye or eyes, training before use is minimal compared to the head-tacked thermal system.

However, the intensifier tube, fiber optic twist, and two lenses are heavy and add significant inertia to the head because of the in-line configuration. From a light detection standpoint, the 3rd Generation cathode has a low peak quantum efficiency, and its spectral cutoff is too short to sense most of the night illumination spectrum. As remarkable as image intensifiers are, it is at least theoretically possible to build something better.

Chapter 3
Generic Staring Imagers

Most imagers today "stare" at the scene and no mechanical scanning mechanism is used. A two dimensional photo detector array is situated in the focal plane of a lens, and that photo detector array is called a focal plane array (FPA). Each photo detector senses (looks at) at a fixed, angular location in the field of view. The array "stares" at the scene, and the imager is therefore called a staring imager.

The essential parts of a staring imager are illustrated in Figure 3.1 The lens at left focuses light on the photo detectors. A point of light in the scene gets spread out (blurred) by the lens. The blurred image of a point source is called the point spread function (psf) of the lens. The light striking each detector generates photo electrons. The photo electrons are integrated (summed), and an electrical signal proportional to incident light energy is generated in each photo detector. The detector signals are collected and amplified.

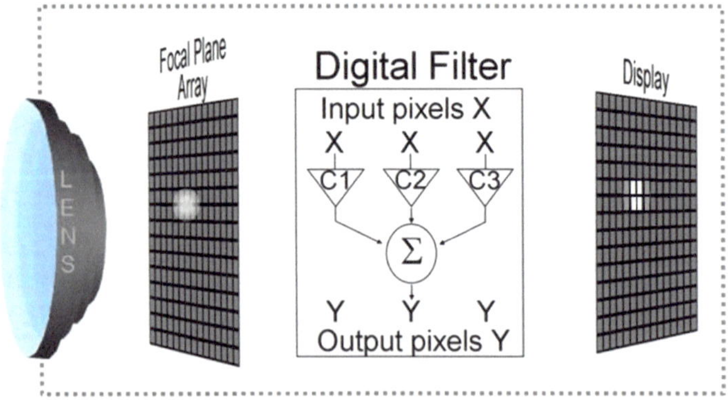

Figure 3.1. A staring imager.

The amplified detector signals are then digitally processed. Digital processing sums and differences near-adjacent pixels in various ways. In the figure, X are input pixel intensities, and Y are digital filter output values obtained by multiplying by coefficients C1, C2, etcetera. Typically, digital filtering enhances contrast and controls gain and level, but it can also reduce noise or improve the displayed picture in other ways. After digital processing, the detector signals are electronically sequenced to light up display pixels that correspond in position to FPA pixels.

The display psf differs from the optical psf. Depending on the location of the light source in the scene, the optical psf can be at any position on the focal plane. However, both the photo detectors and display pixels are at fixed locations. Each photo detector stares at a fixed angular location in the scene. That is, the radiant flux emitted by the scene is sampled.

Figure 3.2 illustrates the effect of sampling. The top pictures A and C show a large blur In A, the blur is centered on a pixel whereas in B the blur is centered on the junction between four pixels. The bottom pictures in A and C show a smaller optical blur. The corresponding display pixel illuminations are shown in B and D. Note that, for a large blur, shifting the optical psf location has little effect on the displayed image, but when the blur is small, the display psf is shift dependent.

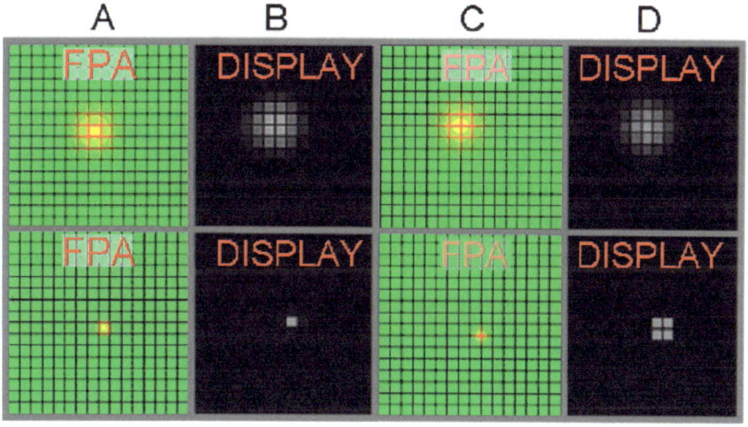

Figure 3.2. In A, light from a point in the scene forms a blurred yellow image on the green array of photo detectors. If the blur is small as shown in the lower picture, then the image of the blur changes depending on the position of the optical psf relative to the photo detectors.

The degradation in image quality from a large optical blur is always worse than the degradation due to aliasing, and eliminating aliasing by purposely designing optics with a large blur is seldom a good idea. However, the fact that blur is bad does not make aliasing good. The images in Figure 3.3 provide an example that large blurs are worse than the removed aliasing, but aliasing still represents corruption of the image.

The picture at left in Figure 3.3 has aliasing, especially the corruption of the thin gun barrow. The picture on the right in Figure 3.3 has all of the aliasing removed. The gun barrow is now straight, but blurry. Clearly, the better picture is on the left. Aliasing can be corrected without blurring the picture by using smaller and more photo detectors to better sample the optical psf.

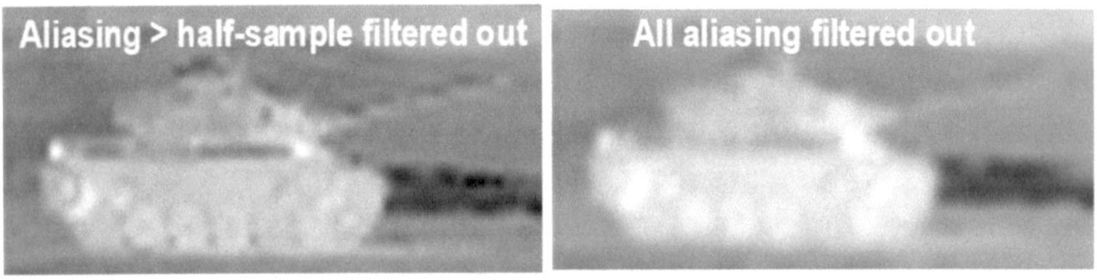

Figure 3.3. These pictures demonstrate that removing aliasing by blurring the image is a mistake, but aliasing does degrade an image.

Aliasing can corrupt an image and is a particular problem for navigation imagers because of the importance of scene texture. Aliasing in an image changes as the position of each scene feature moves within the field of view. Since it is the interplay of scene features with vehicle motion that generates motion parallax, a significant amount of aliasing will corrupt parallax motion cues.

Aliasing is not a problem with either the 1st or 2nd Generation PNVS imagery. The horizontal signal was well sampled in both systems. The 1st Generation imager used interlace, and mechanical interlace overlaps the photo detectors vertically for good sampling. The 2nd Generation system also used mechanical interlace, but additionally, the time-delay-and-integrate detector rows were vertically offset for good sampling even without interface.

Parameters Used to Quantify Imager Performance

This section describes the parameters commonly used to quantify imager performance. Since the book is about imager design, we must use the parameters and methods understood by the industry, and industry invariably uses Fourier analysis to evaluate imager performance. However, the function and utility of Fourier analysis parameters can be understood without delving into Fourier analysis theory. This section only describes parameters cited in later chapters. A thorough discussion of imager Fourier analysis methods can be found in *Analysis and Evaluation of Sampled Imaging Systems* available from SPIE.

Field of View and Imager Magnification

Field of view (FOV) and optical magnification were described in Chapter 1, but a way to calculate those parameters was not provided. After FOV and magnification, radiometric units are described, except the use of temperature as a radiometric unit is discussed in the section on thermal imagery. Sources of noise are not discussed because we do not use noise parameters in later chapters. Finally, the logic and utility of the modulation transfer function (MTF) are explained.

Figure 3.4A shows how to calculate FOV, and B shows how to calculate magnification. A milliradian is the angle subtended by one meter at a range of one kilometer, and there are 17.5 milliradians in one degree. FOV is the length of scene (L) visible in the camera divided by the range from camera to scene.

Magnification is the angle that the display of the object subtends at the eye divided by the angle the object subtends at the camera. Magnification gets bigger if the man places his head closer to the display. Unity magnification occurs when the angle at the eye and the angle at the camera are equal, and unity magnification is used in navigation imagers so that our brains interpret motion cues correctly.

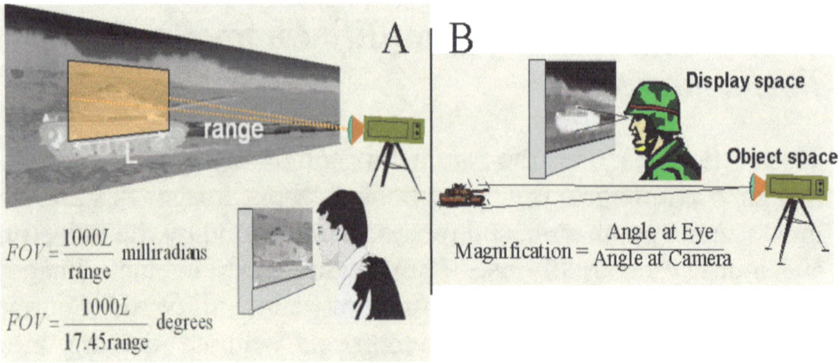

Figure 3.4 showing how to calculate field of view (A) and magnification (B).

Radiometry

An imager purchased today will be used sometime in the future, somewhere in the world, to do some visual tasks in support of some as yet unspecified mission. We want to evaluate how well a navigation imager renders the many objects in a complex scene. We do not want to evaluate range to a specific object.

Figure 3.5 shows a tank, an example specific object, at various ranges from the imager. The line grid outlines the instantaneous FOV for each imager pixel. The number of imager pixels that intercept the tank varies with range, and the total tank signal sensed by the imager is inversely proportional to range squared.

Calculating how well an imager renders an individual object at range is a complex problem, and fortunately, it is not a problem that we need to solve. Instead of selecting an individual object to render with the navigation imager, we follow the normal image analysis procedure and represent the scene as a random, Normal distribution.

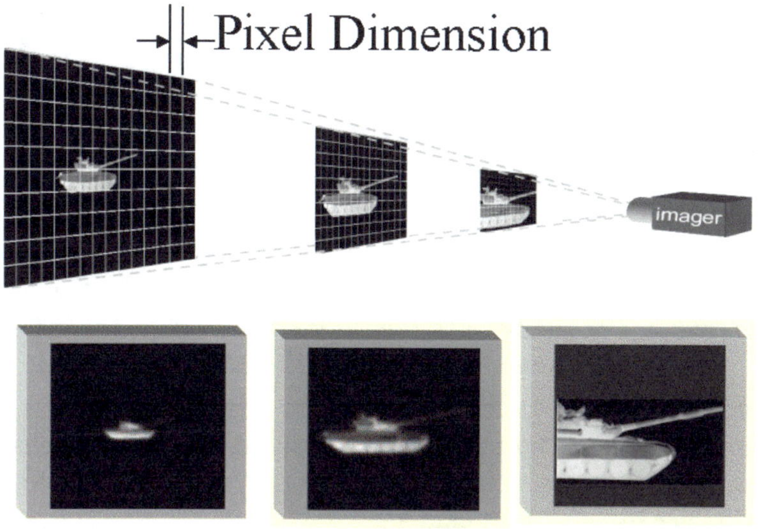

Figure 3.5 illustrates that fewer imager pixels intercept the target as range increases.

We want to evaluate how well the imager renders a complex scene, and the scene is represented by a random, Normal distribution. In the next section, we will describe imager sine wave frequency response. The random, Normal pattern illustrated in Figure 3.6 generates a constant amplitude signal at all spatial frequencies. That is, the random, Normal distribution lets us evaluate how well the imager renders many, diverse objects, and that is exactly what we want to do.

There is another benefit to the random, Normal distribution, because it generates a constant amplitude signal at all spatial frequencies regardless of range. In Figure 3.6, the imager at left (A) is closer to the scene than (B) at right. However, because B is further from the scene, the area subtended by each pixel is larger, and the signal per pixel remains the same as in A. The pixels at left in Figure 3.5 are also larger at longer range, but the specific object (the tank) does not generate a constant amplitude signal at all spatial frequencies.

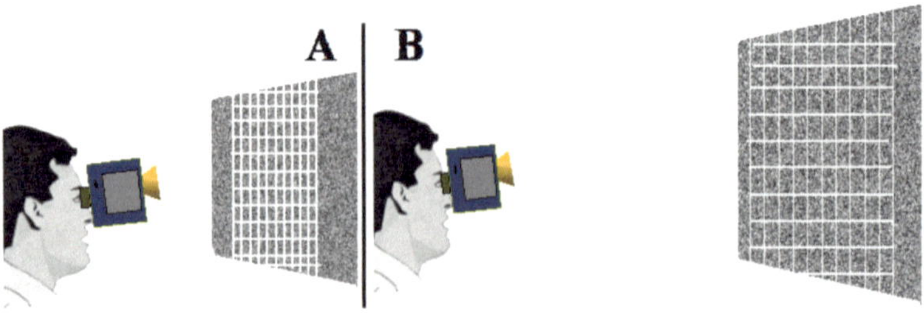

Figure 3.6. An imager views a random, Normal distribution from two ranges. The imager A on the left is closer to the scene but each pixel subtends a smaller area.

Imagers respond to energy, that is Joules, and the scene illuminations and thermal contrasts provided in the next chapter are expressed as range independent Watts (Joules per second) per centimeter squared per steradian.

To summarize this radiometry section, we take the traditional analysis approach and analyze imagers in imager angle space viewing a random, Normal scene. We make those choices, because over years of use, the scene will be spatially diverse, and representing the scene by one rock, one hill, or one tree will not represent actual use. Having decided to evaluate imager performance in imager angle space while viewing a random, Normal scene, terrain contrast can be expressed as a range-independent watts per square centimeter per steradian.

Modulation and System Transfer Functions

Fourier analysis is virtually always used when evaluating imager performance, so it is necessary that later chapters in this book use Fourier parameters to describe imager characteristics. This section starts by describing why Fourier analysis is used, and we then describe the utility and logic of commonly cited Fourier parameters.

Fourier analysis simplifies the mathematics used to analyze imagers. To understand why that is true, consider the various ways in which a camera blurs the image of a scene. Each imager component, the lens, detectors, and display, all sequentially increase image blur. As illustrated in Figure 3.7, the image on the

FPA is slightly blurred because of the lens psf. The photo detectors sum light over the detector active area, thereby further blurring the image. If no digital filtering is used, there will be no digital filter blur, but the display pixels have physical size, further blurring the final picture.

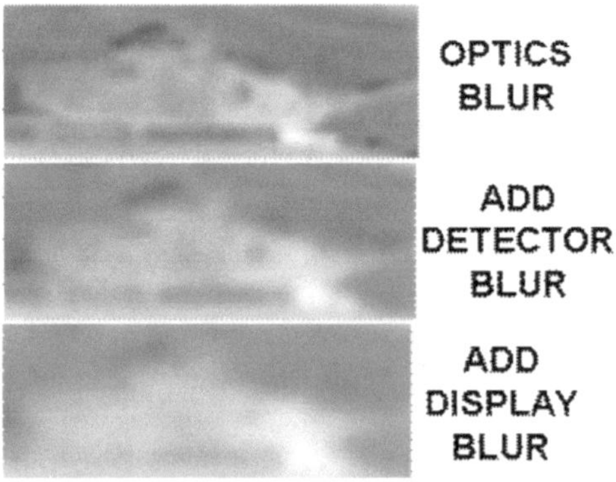

Figure 3.7 illustrating that each imager component adds more blur.

Each camera blur, optics, detector, and display, smears the image. In space, an image smear is a convolution, and convolutions are difficult to calculate. On the other hand, if the images and blurs are treated as sums of sine waves, then each blur is applied mathematically as a product of sine wave amplitudes. That is, the calculations needed to analyze imager characteristics are easier in the Fourier Domain.

Each individual blur is described by its own Modulation Transfer Function (MTF), and the camera blur is described by the product of all the individual MTF. The total camera blur, the product of all the individual MTF, is called the Camera Transfer Function. The remainder of this section describes the utility and logic underlying MTF.

Any spatial pattern, an image or shape, can be described by the sum of sine waves. A sine wave is shown in Figure 3.8A. Part of the figure shows a sinusoidal intensity pattern, and the intensity of the sinusoidal pattern is plotted above. Although the modulation period, the spacing of peaks and troughs, is constant across the figure, the pattern shown is not a single frequency.

One frequency extends throughout space; it has an infinite number of periods.

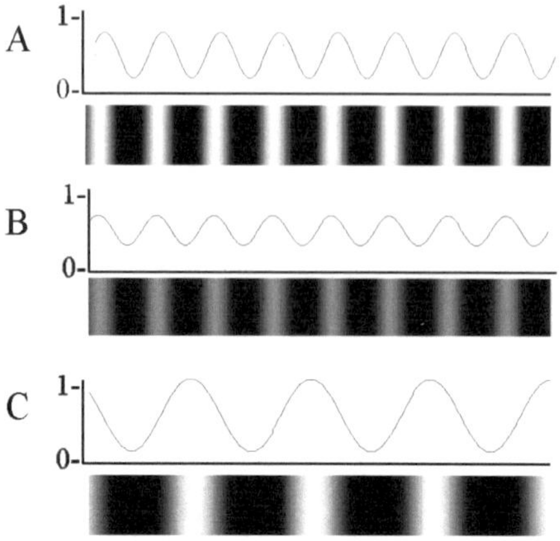

Figure 3.8 shows a sine wave in A, the same frequency but different amplitude in B, and a different, lower frequency in C

Frequency value is defined by the inverse of the period, so if the period is in milliradians, then frequency is in cycles per milliradian (1/milliradian). If period is in meters, then frequency is in cycles per meter (1/meter). Figure 3.8B shows the same frequency as in A but with a different amplitude. Figure 3.8 C shows a lower frequency sine wave, meaning the period (the peak-to-peak or trough-to-trough distance) is longer.

The modulation value for each sine wave is given by the formula MTF equals (High – Low) divided by (High + Low). High is the peak amplitude of each period, and low is the minimum value. A MTF value is calculated for each frequency and amplitude, and a plot of those values versus frequency is called the Modulation Transfer Function (MTF).

In the Fourier Domain, spatial patterns are described by the sum of many sine waves, each sine wave having a different spatial frequency (a different period) and a different amplitude. For example, Figure 3.9 shows the sine waves that are summed to represent a 3-bar pattern. Individual sine waves are at left, and the sum of the sine waves at left is shown to the right. The final bar pattern at the bottom right is constructed using many sine waves not shown, but the gradual appearance of the 3-bar pattern when summing the sine waves at left illustrate the idea.

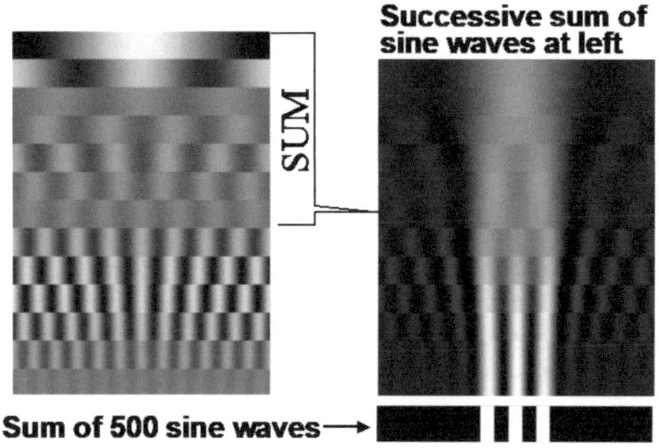

Figure 3.9 shows sine waves that construct a 3-bar pattern.

The system or Camera Transfer Function is the product of all the component MTF (optics, detector, etcetera). The CTF quantifies total camera blur.

Chapter 4
Motion Blur

There are two sources of motion blur. The first is scene motion across the photo detector array, and the second is eye motion across the helmet display. Neither blur is a problem for fielded night vision systems. Image intensifiers update the display virtually real-time. The PNVS thermal imager uses mechanical scanning, and that results in a detector dwell of tens of microseconds. Also, both generations of PNVS use a Cathode Ray Tube (CRT) display, and again the dwell is extremely fast.

Since display blur is associated with natural movement of the eyes within the head, it seems likely that pixel dwell will affect any system using a helmet display. We have psychophysical data on which to base motion blur estimates, and a flight evaluation of the blur estimates supports those blur predictions.

Predicting the blur due to scene-detector motion is relatively easy for the system described by Figure 1.3, because the imager is fixed in vehicle coordinates. In the other navigation imager configurations, Figures 1.1, 1.2, and 1.4, the motion blur changes as the operator moves his or her head. While head motion might decrease blur in some cases, it will increase blur in other cases, and only fixed line-of-sight motion blur is considered in this chapter.

Blur Due to Display Pixel Dwell

When a person moves his head to, say, read a roadside sign as he speeds past it, his head and eyes work together in a natural and involuntary way. Moving the head is voluntary, but it is the involuntary coordination of head and eye movements that allows the driver to read the sign.

A helmet display is attached to the helmet which is attached to the head. In one situation, perhaps the eye moves within the head to fixate a moving object, and the head moves to keep the object sufficiently centered for eye tracking. Or perhaps the person moves his head to look for a different direction of travel, then the eye saccades to find interesting locations. The eye moves within the head, and blur occurs as the eye's fixation point moves across the display pixels.

Based on published psychophysical data, humans have full visual acuity with eye movements of at least 30 and up to 60 degrees per second relative to the head. Since no loss of vision is associated with a rate of 30 degrees per second, one might assume that rate occurs when engaged in eye-head tracking of a moving object, or eye-display blur might occur when just using the head to look around.

Figure 4.1 at left shows the helmet mounted camera used in a flight evaluation of blur due to display pixel dwell. The aviators were flight qualified on the Apache PNVS, but the bi-directional scanning on the 1st Generation system made it unsuitable for this purpose. Also, the head-mounted camera eliminated turret lag. A black-out cloth was draped over the visor so that outside vision was obscured. The flight tasks and evaluation methodology are described in Chapter 7.

The graph at right in Figure 4.1 shows the modulation transfer functions (MTF) of the helmet-mounted camera, the AHP, and three motion blurs. The AHP was not part of the evaluation, but it is a stand-in for fielded systems that have a better MTF than the camera used for these flights. The three motion blur MTFs represent 4, 8, and 16 millisecond dwells and a 30 degree per second motion.

The 4 millisecond blur did not affect flight control, workload, or meeting the standard for each flight task. The 8 millisecond blur degraded the imagery, reduced confidence, increased workload, and resulted in poor flight control as evidenced by failure to meet the flight standards. The 16 millisecond dwell degraded the imagery to the point that the camera image was deemed unacceptable for flight control.

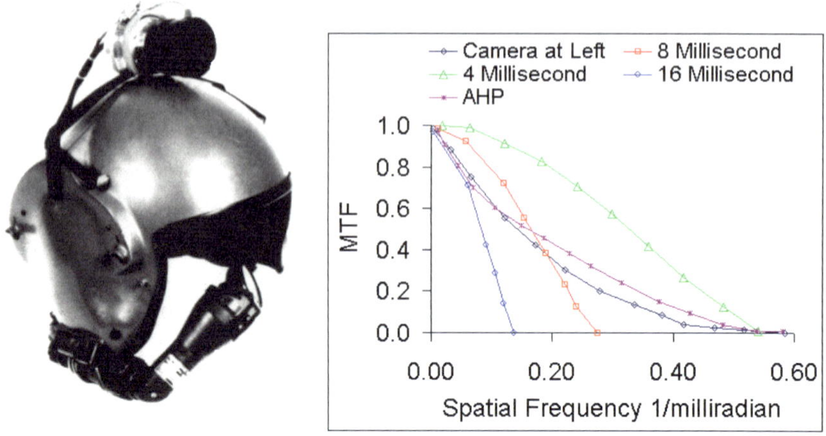

Figure 4.1 shows the MTF of the helmet camera and of three motion blurs.

The flight results are consistent with the plotted MTF in Figure 4.1. In other words, it appears that assuming a 30 degree per second residual motion while head-tracking the scene provides a realistic assessment of the impact of display pixel dwell. Helmet mounted display pixel dwell should be limited to no more than 4 milliseconds. If a future camera improves on the AHP MTF, then a four millisecond display pixel dwell might become a limiting MTF for that future camera.

Video motion appears smooth to us, not because there are so many frames displayed that we cannot see each little jump, but because of something that psychophysicists call *smooth apparent motion*. The eye can fixate with full acuity motions greater than 30 degrees per second, and one standard video frame or field takes 1/60th of a second. A jump of more than half a degree would be quite visible if presented statically. However, when presented as a continuous sequence, the portrayed motion appears to be smooth, covering all positions across the transited route.

Motion Blur Due to Photo Detector Dwell

If we assume that the imager line-of-sight is fixed in vehicle coordinates, then the motion blur is easily calculated using the formula below. To simplify the discussion, the vehicle velocity vector is coincident with the FOR coordinate origin. Motion blur (B_m) in milliradians for a dwell (D_w) in milliseconds depends on a simple relationship between vehicle speed (V) in knots per hour (knt), the range to the stationary object (R_{ng}) in meters, and field of regard (FOR) angle (α) in milliradians.

$$B_m = (.0085) * V * D_w * \sin(\alpha)/R_{ng}.$$

The graph in Figure 4.2 requires some work to interpret, but there are too many speed, range, angle, and detector dwell times to present individual results. The graph in Figure 4.2 assumes that vehicle motion is straight ahead, the photo detector dwell is one millisecond, and the several lines are for different, labeled FOR angles. The ordinate is blur in milliradians, and the abscissa is the ratio of speed in knots per hour to viewed object distance in meters. The blur for longer or shorter detector dwells is obtained from the graph by multiplying the ordinate milliradian blur by the new detector dwell in milliseconds.

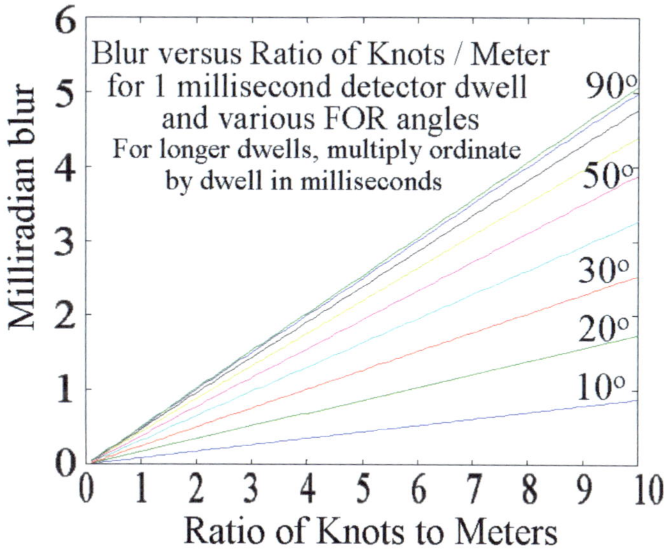

Figure 4.2 showing the blur in milliradians associated with various speed to range ratios. The several lines are for different FOR angles. These blurs are for a one millisecond detector dwell.

For example, referring to the NOE picture at right in Figure 4.3, assume the pilot is looking at a 40 degree angle to the right of the aircraft, that the vehicle speed is 30 knots, and that the trees are ten meters to one side. Then the speed to range ratio is three, and the blur for a one millisecond detector dwell at the 40 degree FOR angle is one milliradian. If the detector dwell is four milliseconds, then the motion blur would be four milliradians. A four milliradian blur would severely degrade the image.

Figure 4.3. The picture at left shows contour flight and the right-hand picture shows Nap of the Earth.

Using the left hand picture in Figure 4.3, if the aviator looks 50 meters ahead at 30 degrees to the right to find a better flight path for 100 knots per hour contour flight, the four millisecond dwell blur would be two milliradians. That would degrade flight control.

The calculations presented here strongly suggest that a 4 millisecond dwell will result in unacceptable motion blurs if the navigation imager line of sight is fixed in vehicle coordinates. For navigation imagers represented by Figures 1.1, 1.2, and 1.4 that use head-tracked imagers, calculating motion blur is complicated, However, assuming that the aviator can compensate for excessive blur using head motion is an untested hypothesis that is likely wrong. At the least, that hypothesis should be verified by either a flight or ground vehicle evaluation.

In summary, a four millisecond photo detector dwell is unsuitable for navigation imagers illustrated by Figure 1.3 where the imager is fixed in vehicle coordinates, and it is highly likely that a four millisecond dwell is unsuitable for any navigation imager. It is not apparent from the graphs in Figure 4.2 that even a one millisecond dwell is suitable for navigation imagers.

Chapter 5
Thermal Imager Design

This chapter describes the benefits of operating in the long wave infrared spectrum. This chapter also provides the thermal contrast levels where fielded systems provide adequate imagery. Also, although staring imagers have a significant signal to noise advantage, some of the benefits of mechanically scanning a linear photo detector array are described.

There are, of course, many factors that affect the final imager design choice, like cost, system weight and power, the several problems associated with selecting a cryogenically cooled photo detector, and more. All of those factors weigh against performance. Cost versus performance trades are necessary. The goal of this chapter is to provide insight on the performance side of those trades.

The first thermal section describes the fundamental problem that must be solved when designing any photon-counting thermal imager. By photon-counting we mean not uncooled. Since this chapter is about fundamental design considerations, we use indium antimonide (InSb) and mercury cadmium telluride (HgCdTe or MTC) cooled to 77 K to represent mid wave and long wave infrared detectors. Cryogenic InSb and HgCdTe are not selected because one will be the photo detector of choice for advanced navigation imagers, but because they are the best representatives of detectors in their respective spectral bands.

Once the fundamental problem with thermal imaging is described, data and analysis are presented to explain why long wave infrared provides critical advantages over mid wave for navigation imagery.

The final thermal section describes why a mechanical scanner has a means to improve sensitivity that is not available for staring arrays. In theory, a staring imager has such a big advantage over a linear array that many performance compromises can be made and the staring imager will still outperform the linear array. In practice, that might not be the case.

Uncooled thermal imagers are not discussed in this chapter, but they are the only thermal imaging technology considered in Part 2 of the book. Part 1 of the book describes design criteria, and the operating principles of uncooled imagers are too different from photon counters to make their inclusion here useful. The goal of this chapter is to explain why the long wave signals are stronger than mid wave, and using photo detectors that sense using physically different mechanisms would not benefit that discussion.

The Problem Faced by All Thermal Imagers

Everything radiates thermal energy. At first look, that might seem to be a good thing, because nothing is invisible. But the fact that everything radiates, and generally radiates about the same amount of energy, is not a good thing when designing an imager. The fact that everything radiates, and radiates by about the same amount, means that thermal images are always very low contrast.

Figure 5.1 shows a thermal image of a tank. Picture (a) at top left is an image of the light received by the camera. In Picture (b) at bottom left, the display black level is set to the minimum luminance in Picture (a). Thermal cameras remove the bright intensity pedestal shown in Picture (a), but the pedestal is always present. The plot at right in the figure is the intensity along the dashed vertical line in Picture (a). The dashed line is continued down through Picture (b) to show that the line intersects the hot areas on the tank.

The bright area on the back, top of the tank is heat from the engine, and the tracks are hot because the tank had been moved from the parking area to an open field for the picture. In other words, many parts of the tank are hot to the touch, but even those radiate little more thermal energy than the grass beside the tank.

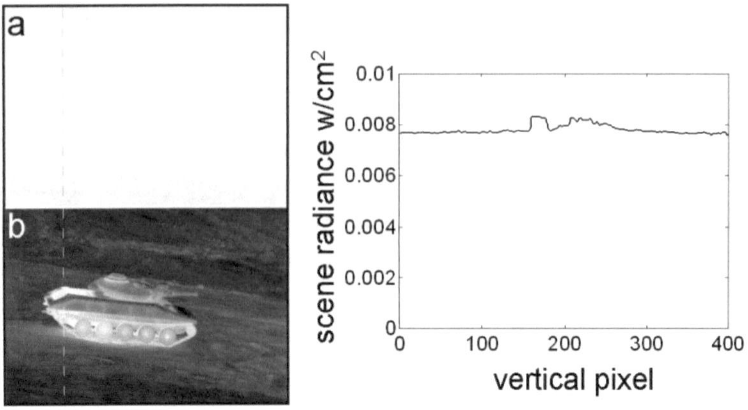

Figure 5.1 illustrating that infrared images are low contrast.

We live in a 300 K world, and we judge hot or cold relative to room temperature. Objects radiate heat, however, based on their absolute temperature. In an absolute sense, the hot engine is not that much hotter than the grass or dirt under the tank tracks.

When a photo detector is exposed to light, the number of photo electrons generated each second is proportional to the watts hitting the detector. However, even if the light level is constant, detector noise will be proportional to the square root of the number of photo electrons. Since all thermal scenes are low contrast, scene details are represented by a small change in photo electron number, but noise is proportional to the square root of the large signal generated by the 300 K background.

Although the signal is small and the noise is large, the signal is constant but the noise is random, so integrating (summing) the photo detector signal for a long time will eventually raise the signal above the noise. In other words, thermal imagers require a long exposure time to render good imagery.

Long Wave Versus Mid Wave

In theory, the 8 to 12 micron long wave spectral band is optimum for terrestrial thermal imaging, In theory, a long wave imager reaches a specified signal to noise ratio more than ten times faster than a mid wave imager. That factor of ten decrease in dwell time (exposure time) is critical in scenarios where motion blur is a problem. In theory, the long wave provides a factor of 3.5 in signal to noise improvement when the dwells are equal. That sensitivity advantage is critical in scenarios with poor scene contrast. In theory, the long wave outperforms the mid wave in any terrestrial environment.

The fact that long wave and mid wave cameras tend to have the same laboratory-measured sensitivity is often cited to show that mid wave and long wave perform the same. However, it is also a fact that the long wave cameras achieve the same sensitivity ten times faster than the mid wave, and that information is not recorded with the sensitivity data. The reason for laboratory equivalency of the two spectral bands will be explained.

As discussed in the previous section, in order to achieve good sensitivity, each infrared photo detector integrates photo electrons on a capacitor. Since each detector in a two dimensional array has its own capacitor, the capacitors are small. Using existing methods and materials, the number of electrons that can be stored is limited. Since that capacitor limitation applies equally to both mid wave and long wave imagers, they store the same number of photo electrons and therefore exhibit the same sensitivity.

Note, however, that the long wave imager achieves that sensitivity ten times faster than the mid wave. In the long wave case, the electron well fills quickly and signal integration must stop, whereas the mid wave continues to integrate until it also fills its electron well.

The capacitor technology limitation obscures the innate advantage of long wave for terrestrial imaging. With the same capacitor, long wave achieves the same sensitivity as mid wave, but ten times faster. On the other hand, there are several on-going developments that will likely fix the capacitor storage problem, and the long wave will then have a factor of 3.5 sensitivity advantage over the mid wave.

To be clear, using current technology, the long wave has a significant, factor of ten, advantage in scenarios where motion blur is a problem, but the mid and long wave perform the same under very poor contrast conditions. When the capacitor storage problem is fixed, the long wave will provide a factor of 3.5 better signal to noise under poor contrast scenarios. The benefits of long wave over mid wave are quantifiable and significant.

Mid Wave and Long Wave Thermal Contrasts.

Aside from one situation, thermal contrasts are essentially the same in the two spectral bands. The one situation arises because the long wave sky is cold and the mid wave sky is not. That means that terrestrial objects viewed against the sky will be highlighted in the 8 to 12 micron spectral band. See Figure 5.2. For objects viewed against the sky, the long wave contrasts are 1.8 to 5.3 times higher than in the mid wave. For non-horizon viewing, the contrasts are essentially equal. Whether that particular phenomena is important in a practical sense is not known. Intuitively, it seems like enhancement of horizon contrasts would help pilotage, but we have no user feedback to support that assumption.

Picture # horizon	1	3	5	7	9	11	13	15	16	18	
Ratio LWIR/MWIR	2.5	2.6	3.5	2.4	5.3	2.8	1.8	2.3	2	3.5	
Picture # non-horizon	2	4	6	8	10	12	14	17	19	20	21
Ratio LWIR/MWIR	1	0.8	1.1	0.9	1.1	0.9	1.1	1.7	1.1	1.2	1

Figure 5.2. Examples comparing long wave to mid wave thermal contrasts.

Figure 5.3 is a picture taken with an uncooled camera. Branches viewed against the sky are brightened by the cold background in the long wave infrared.

Figure 5.3. Uncooled thermal picture of tree line showing the highlight effect.

Forty years after the 6th Cavalry became the first unit equipped with Apache Helicopters, terrain thermal contrast data is still almost non-existent. Some data were collected during an early 1st Generation PNVS evaluation in Germany during November 1981. That evaluation compared flights using PNVS to same night flights using early, prototype versions ANVIS image intensifiers. The image intensifiers outperformed the PNVS, and the Tree-to-Ground thermal contrast data collected at the same time as the flights explains why.

The contrast data is shown in Figure 5.4. Based on the histogram data in the figure, the early PNVS imagery would have been poor to very poor half the time and just noise the rest of the time. The good news from those flights was that ANVIS performed well, and that 1st Generation PNVS sensitivity was almost good enough to operate in a very poor thermal contrast environment. That flight evaluation gave impetus to the very successful fielding of 3rd Generation image intensifiers, and also provided technical evidence that the then available time-delay-and-integrate detector technology would solve the PNVS sensitivity problem.

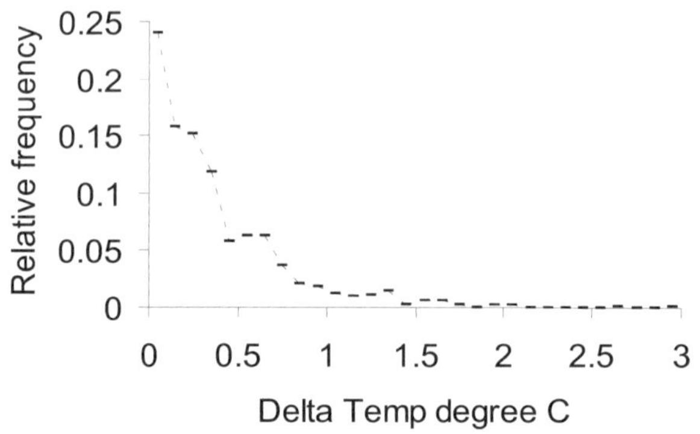

Figure 5.4. Histogram of tree-to-ground thermal contrast taken at night in Germany in November. "Tree-to-Ground" is the standard deviation of an image captured by an imaging radiometer.

In summary, the histogram in Figure 5.4 and the non-horizon data in Figure 5.2 apply to both mid wave and long wave thermal imaging. The two spectral bands are also equivalent in terms of optical blur, because with wide FOV optics, aberrations dominate the blur, and diffraction effects are small. In terms of imager resolution under good contrast conditions, any differences between mid and long wave optical point spread function result from optical material choices and not fundamental physics.

However, long wave thermal outperforms mid wave thermal in two significant ways. First, under good contrast conditions, the long wave motion blur will be a tenth of the mid wave blur. Second, under poor contrast conditions, the long wave signal to noise is 3.5 times better than the mid wave.

Some Benefits of Mechanically Scanned Linear Detector Arrays.

The benefit of staring technology relative to a linear array of photo detectors is normally calculated as follows. If the staring array has 2K by 2K detectors, then the signal to noise improvement over a linear array would be $\sqrt{2000}$ or 45. That is a huge signal to noise improvement, and it appears to suggest that there is plenty of room to consider hot detectors, slower optics, and short dwell times.

The problem with that comparison is that it assumes that the linear array of detectors is limited by the same producibility and fabrication limitations as the staring array, and that is far from true. A lot of flexibility accrues from using a linear TDI array of detectors. The Advanced Helicopter Pilotage TDI detector is used in an example. AHP is compared to an arbitrarily selected staring array to illustrate some design considerations.

The graph at left in Figure 5.5 shows the spectral response of InSb and MCT (HgCdTe); those detectors represent the best performance in their respective spectral bands. In both cases, the short wavelength sharp cutoff is the result of a cold filter. At right in the figure, the graph plots 300K background photon flux and the photons from a 1 K signal difference at 300K. The hatched area represents a CO2 spectral cold filter. In that mid wave spectral region, no signal penetrates the atmosphere, but the atmosphere radiates, so all light in that spectral region is blocked.

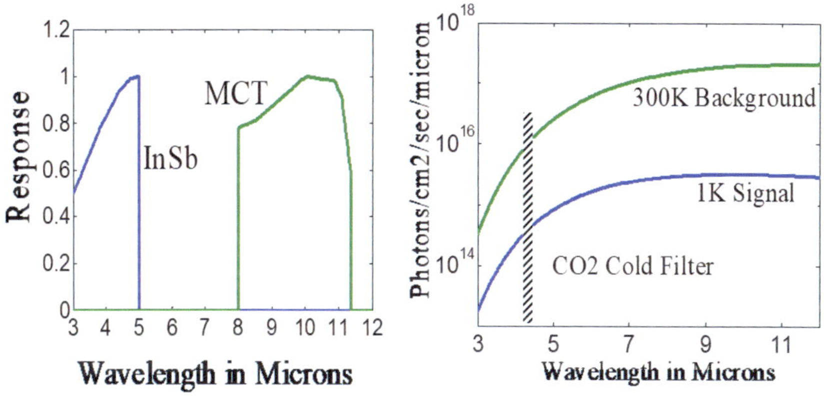

Figure 5.5. The plots at left are the spectral responses of InSb and HgCdTe. At right, the graph plots photon flux from the background and from a 1K signal.

AHP uses the HgCdTe, cryogenically cooled, photo voltaic diodes with the wavelength cutoff shown in Figure 5.5. The replacement imager is assumed to be a 25 micron InSb operated at the full 16 millisecond dwell time. Assume that the two imagers use the same F/number optics.

The AHP dwell is 24.4 microseconds, and that results in a factor 25 reduction in signal to noise compared to the InSb camera.

However, the InSb operates in the mid wave spectral band, and the detector area is 11.2 times smaller than AHP, so the factor of 25 reduction becomes $25/(3.5 * \sqrt{11.2})$ which equals 2.1. In order to avoid the cryogenic cooler, we will replace the InSb with nbn and call the sensitivities equal.

So far, none of our assumptions are pessimistic towards the mid wave imager. The InSb photo detector size might be considered pessimistic, except making the photo detectors larger would create producibility problems. One might argue that replacing a mechanically scanned imager that uses cryogenic cooling with a hot, producible, 25 micron staring array is sufficient benefit, and that improving sensitivity of the current system is not necessary.

Consider, however, that the new imager operates with a 16 millisecond dwell, and the 16 millisecond dwell would certainly prove unsuitable for a navigation imager due to motion blur. Also note that the two samples per dwell both horizontally and vertically of the older system is now one sample per dwell in both directions, and that causes aliasing because of the fast optics used for navigation imagers. Next, consider that the linear array has almost perfect uniformity horizontally, and that vertical uniformity is corrected real time. It is very unlikely that the nbn array can achieve the same level of image uniformity.

This example proves nothing about mid wave, InSb, or nbn; it was simply a possible design example. The example was presented to demonstrate that the older technology had some benefits that, to some extent, offset the sensitivity advantage of new technology staring arrays.

Chapter 6
Reflective Imager Design

This chapter provides night illumination and terrain contrast data for the near infrared (NIR) and short wave infrared (SWIR) reflective light spectrums. This chapter also discusses some of the difficulties with replacing image intensifier technology with solid state staring imagers.

Night Illumination and Terrain Contrast Data

On a moonless night, the primary source of natural illumination in the near infrared (NIR) and short wave infrared (SWIR) spectral bands is airglow. Most of the airglow energy that reaches the ground is emitted by an optically thin layer of excited hydroxyl radical that is about 8 kilometers (km) half-width thick and is centered near an 87 km altitude.

Thus, ground illumination depends on the solid angle subtended by open sky at the viewed location. Most of the light imaged by night sensors is emitted by airglow, not the stars, and airglow is emitted by a high altitude, thin layer of atmosphere, so illumination is emitted by all areas of the sky.

That is not to suggest that airglow illumination is constant or reliable; quite the contrary. However, we will use an airglow illumination level that is considered low enough to represent the low ebb of airglow cycles at locations where airglow is weakest, that is, far from the equator.

If you use an image intensifier to look at the night sky, you will see stars and not airglow. Stars are a point source of light, they are bright at one spot in the sky, whereas airglow emissions are diffuse. The airglow is not visible when looking at the sky, but the totality of light hitting the ground is almost all from airglow.

Because the airglow layer is optically thin, meaning we can see through it, more light is emitted to the side than down. A thin layer is thicker when looked at from an angle other than normal.

Airglow spectral emission is described by L(λ,γ) where λ is wavelength in microns (μ) and γ is zenith angle in radians. Airglow zenith spectral radiance *L(λ,0)* is shown in Figure 6.1, and zenith angle dependence is given by the van Rhijn function

$$L(\lambda,\gamma) \Big/ L(\lambda,0) = \frac{1}{\sqrt{1-0.9733\sin^2(\gamma)}}.$$

Ground illumination is calculated by integrating light from all angles of open sky. Zenith intensity is given by Figure 6.1, and the intensity for all other sky angles is calculated using the van Rhijn Equation to adjust the zenith intensity for non-zenith angles.

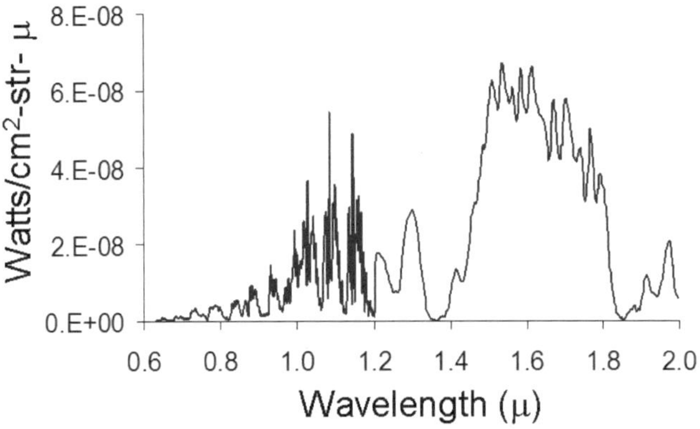

Figure 6.1. Spectral emissions of airglow at zenith.

It is actually quite hard to find a dark night when airglow is the only source of ground illumination. Even thin cloud cover will reflect light from towns miles away and provide much more illumination than "starlight." Nonetheless, there are occasions where the only significant source of illumination is airglow, and that illumination can be reduced by a factor of ten by cloud cover.

The other natural factor that affects night viewing is the reflectance, and especially the variation of reflectance, of ground objects. Figure 6.2 shows reflectance versus wavelength in the near and short wave infrared for 21 types of paint, 41 types of cloth, 18 types of ground (gnd), and 16 types of vegetation (veg).

The standard deviation (std) for each type of material is also shown. Based on this figure, average scene reflectance is 0.5 and the standard deviation in the near infrared (NIR) spectral band is 0.17.

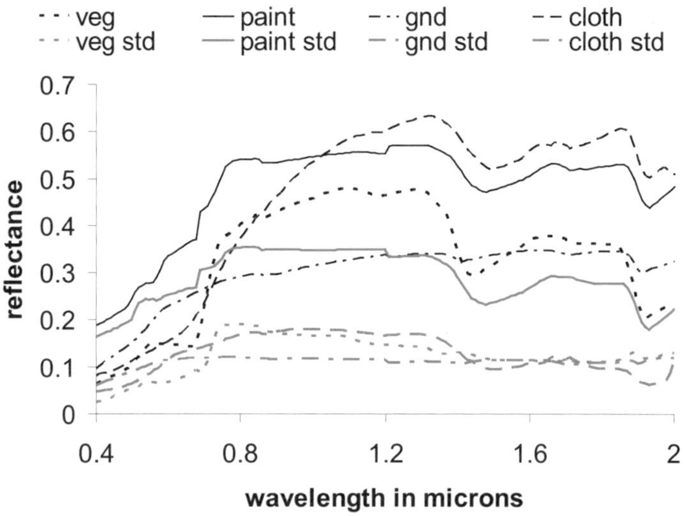

Figure 6.2. Reflectances for various types of material in the near and short wave infrared. The lighter lines show the standard deviation (std) of reflectance for each material group.

Staring Array Replacement for Image Intensifiers

Image intensifiers are heavy because of the amount of glass. Other than that, they have many functional advantages as night vision aids. The image sensing and display are integral to the image intensifier tube. Digital image processing is not necessary to obtain sharp images with high dynamic range, so no external electronics are needed. They operate for hours on a single AAA battery. And they make a dark scene up to 100,000 times brighter.

Figure 6.3 below compares the components of an image intensifier at left with those of a staring imager replacement at right. The staring imager has the potential, in theory, of better night performance, and that is discussed in this section. Also, we can expect that a solid state imager would be both smaller and lighter than an image intensifier.

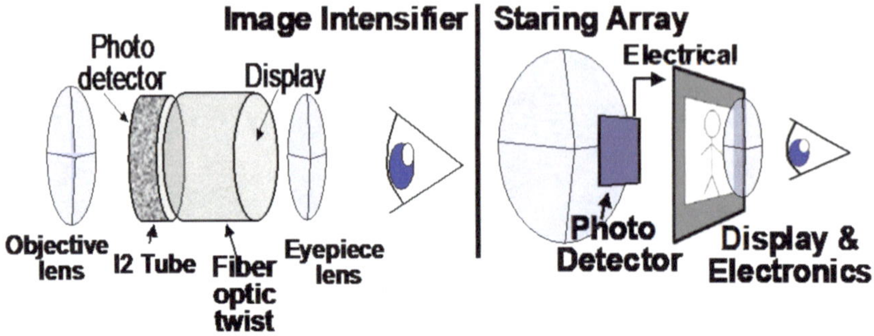

Figure 6.3. At left is the layout of an image intensifier. At right is a typical layout for a staring array goggle

With the imager intensifier, light focused on the image intensifier cathode generates photo electrons that are amplified by the tube and form a bright image on the phosphor screen. That image is inverted by the fiber optic twist, and the eyepiece forms a virtual image of the scene.

The staring device focuses light on a two dimensional array of photo detectors. The electrical signals from the photo detector array are processed by electronics and then displayed. The signal can, of course, be remoted; the imager need not be head-mounted.

One problem with fabricating a head-mounted staring imager is that, except for a CRT, a suitable display does not exist, but that problem will likely resolve soon. Another practical problem with the staring sensor is that the power usage far exceeds that of an image intensifier. The real problem, however, is performance, although the amount of illumination energy available for the staring sensor far exceeds that available for image intensifier. See Figure 6.4 and Table 6.1.

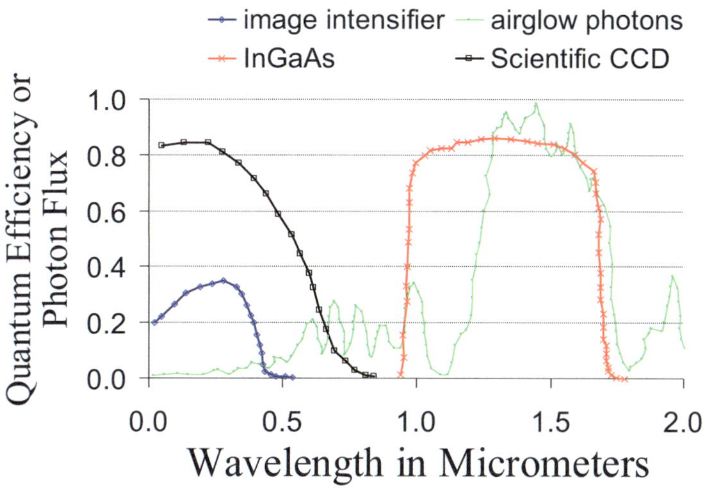

Figure 6.4. QE of 3rd Gen. cathode, silicon, and InGaAs, plus photon flux normalized to 5.4E11 photons per (str-cm²-μm).

Table 6.1. Silicon and InGaAs parameters.

Characteristic	**Silicon**	**InGaAs**
Dark current @ 25C (picoamp/cm²)	6	6,000
Read noise (electrons/pixel/frame)	2	2

The graph shows wavelength versus the quantum efficiency (QE) of 3rd Generation image intensifiers, indium-gallium-arsenide (InGaAs), and the thick silicon layer of a back illuminated scientific charge coupled device (CCD). The graph also shows the photon flux from airglow. The photon flux is normalized to 5.4E11 photons per steradian-centimeter²-micrometer (str-cm²-μm).

The graph demonstrates that there is substantial more light emitted by airglow than sensed by image intensifiers, partly because of the low image intensifier quantum efficiency, but also because most airglow emissions occur at longer wavelengths than sensed by the 3rd Generation image intensifier cathodes.

The problem with using the increased airglow photon flux at longer wavelengths is that photo detector dark current increases markedly at longer wavelengths. Dark current and read noise for the thick silicon used in back illuminated, scientific, charge coupled devices (CCD) and for dark current optimized InGaAs are given in Table 6.1. High resistivity silicon with low dark current and two electron read noise provide the best staring array signal to noise demonstrated to date. However, various non-uniformity problems resulted in poor imagery even with 16 millisecond dwell times. In other words, image intensifiers are more practical, available, and perform much better than any available solid state, staring array imager.

The graph in Figure 6.4 shows that, in theory, a staring imager with better QE and/or a broader spectral band can exceed the performance of image intensifiers and be lighter and smaller. However, at this time, that is theory. Even with a 16 millisecond dwell time, airglow illumination is too dim for existing, room temperature, staring technology to provide a good image. Cooled silicon imagers operated with long dwell times do provide night capability, but the power, weight, size, and motion blur make those devices unsuitable as image intensifier replacements.

Remotely Viewed Image Intensifiers

Figure 6.5 shows three ways that image intensified imagery can be remoted to a display monitor. The first configuration at left uses a fiber optic reducer to connect the large phosphor screen to a smaller, commercially available silicon imager. The second configuration at center accelerates the cathode electrons to a silicon CMOS imager where electron acceleration provides signal gain. The third configuration at right incorporates a microchannel plate (MCP) to provide electron multiplication and gain control. See Chapter 2 for details on MCP operation.

Each of the three configurations use a silicon imager to remote the intensifier image, and that adds signal storage and therefore motion blur. However, the silicon imagers operate at a high signal level, so dark current and read noise and non-uniformity become less of a problem.

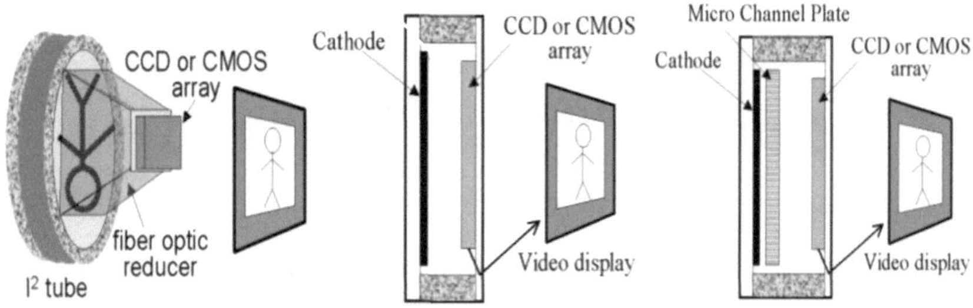

Figure 6.5 showing three ways to remote an image intensifier image to a video display.

Several versions of the left hand configuration were flight tested and performed well based on a subjective viewing of the imagery, but no quantitative data were taken. A version of the center configuration has been incorporated on the Apache Helicopter, and that suggests good performance.

Summary Comments

There are benefits to a short wave infrared solution operating beyond one micrometer that are not obvious from the graph in Figure 6.3 or Table 6.1. Short Wave radiation penetrates fog and clouds better than visible and even near infrared. However, from a practical engineering standpoint, a short wave imager capable of passively viewing under starlight illumination seems even more remote in time than a near infrared device.

The 3rd Generation image intensifier has been in the field long enough and with a broad enough distribution that experience has been used to refine device design. That is not true for the center and right hand configurations in Figure 6.4. However, those configurations are certainly lighter and more compact, and those are important virtues in many applications.

Chapter 7
Field of View and Resolution

At any one level of imaging technology, field of view and resolution are trades. A wider field of view requires more pixels to maintain the same image resolution. A wider field of view requires a superior optical design to maintain the same imager resolution. A wider field of view requires a better helmet display, with more pixels and better optics, to maintain the same imager resolution.

The idea that large fields of view are desirable is certainly correct, but concluding that achieving a large field of view can be at the expense of image resolution is not supported by any pilot survey or flight evaluation. Bigger fields of view are definitely desired by users, but not at the expense of image resolution.

The first section of this chapter summarizes user surveys on field of view and resolution trades. The subsequent section describes flight evaluations on the subject.

User Surveys

User surveys were conducted in 1987, 1990, and after Desert Storm in 1992. These surveys relate only to the imaging device. An aviator will change the aircraft airspeed, altitude, and flight profile as needed to adapt to the conditions encountered. As night sensor imagery degrades, the pilot will also depend more on the instruments and the Head's Up Display symbology. The engineering trades for a night vision sensor relate to the ability of the sensor to deliver the desired visual information; those trades do not relate to the ability of the entire weapon system to accomplish a mission.

Note that, during the years that surveys were conducted, the fielded thermal imager was a 1st Generation device, it was not the currently fielded system. When considering the aviators survey answers, refer to the 1st Generation imager characteristics in

Chapter 2. The survey answers on the old system are included here because they represent field experience with a physically characterized system. Remember, however, that to the aviator, poor resolution equates to poor imagery and not pixel count. The better sensitivity of the modernized PNVS will definitely change survey answers.

When there is good thermal contrast in the scene, and in the absence of fog, heavy rain, or snow squalls, the PNVS thermal imager supports terrain (NOE and contour) flight. Good thermal contrast occurs when there has been clear weather with sunshine for at least a few hours during the day, heating up objects in the background scene. If there has been no sunshine during the day, or if there has been only a little sunshine followed by heavy rain or hours of drizzle, the thermal contrast will be poor, leading to poor visual flying conditions.

Further, the thermal radiation which PNVS images is attenuated by heavy fog and by the atmospheric water vapor content found with heavy rain and persistent drizzle. Image contrast might be poor even when the scene is providing a usable thermal signature. Thus, poor local weather, such as patches of fog or squalls, may make terrain flight difficult at the midpoint of a flight, even though conditions are good at the starting point and destination.

ANVIS performs well under clear starlight conditions but becomes marginal to unusable under overcast starlight conditions. Heavy fog will shut down ANVIS. Even a moderate fog can severely degrade imagery if flight is toward the moon; scattered light from the fog can severely degrade contrast and mask the view of the terrain. Also, ANVIS tends to "bleach out" or shut down when bright lights are in the field of view; this occurs around city lights, when flying toward the moon if it is low on the horizon, and under dawn and dusk conditions.

Flying over land that has no features, such as the sand dunes of Saudi Arabia, presents a challenge; judging distance and closure to the ground requires scene detail. Areas devoid of distinguishable features, such as snow fields, lakes, and dry lake beds, will provide poor imagery for terrain flight. Under these circumstances, the availability of flight symbology is critical.

Pilots express strong feelings that thermal sensors and image intensifiers are complementary and that both are needed for night contour and NOE flight. The combination supports flight under a wider range of conditions than either alone, although environments certainly exist where even the combination will not support terrain flight.

Also, each sensor brings a unique capability to the aircraft. The two sensors operate in different spectral bands and depend on different physical principles for performance. The ability of the aircrew to detect wires and other obstacles is significantly enhanced. Even on poor thermal nights, the PNVS provides a good capability to perceive and react to close-in targets. Even on nights with poor illumination, ANVIS gives the ability to see town lights and therefore provides navigational aid; because ANVIS can see aircraft running lights, it also provides a better ability to fly formation as well as safety from collision with other aircraft.

In each of the three surveys taken between 1987 and 1992, the aviators were asked to answer questions, based on their total flight experience, about needed design improvements in field of view and resolution for ANVIS and PNVS. In an operational context, sensor resolution refers to image quality and therefore depends on the sensor sensitivity as well as the optical resolving power of the sensor.

The results of all the surveys are consistent and can be summarized as follows. Based on total flight experience, pilots rate both the FOV and the resolution of ANVIS as acceptable. Pilots would choose to expand ANVIS FOV but not at the expense of current image quality. On the basis of total flight experience, pilots rated the PNVS FOV as adequate but the resolution as inadequate; they would improve image quality before expanding FOV. The pilots are interested in increased PNVS FOV but only in combination with improved image quality.

A summary of the responses to each survey is given below.

The 1987 survey queried 49 Apache helicopter pilots, all with PNVS thermal imager experience; 29 of these aviators had

ANVIS experience. When given an open choice of which sensor they preferred, 42 of 49 wanted both PNVS and ANVIS.

The Apache crews were asked to give an overall rating for PNVS and ANVIS as to adequacy of FOV and resolution (image quality); they were to answer based on their total flight experience. Table 7.1 summarizes how many pilots rated FOV and resolution as good, adequate, and inadequate. In general, the pilots rated the PNVS FOV as adequate but the resolution as inadequate. They rated both the FOV and resolution of ANVIS as adequate.

Table 7.1.

Sensor/Feature	Good	Adequate	Inadequate
PNVS FOV	5	35	9
PNVS Resolution	1	18	30
ANVIS FOV	9	17	3
ANVIS Resolution	13	13	3

The large majority of Apache aviators, 45 out of 49, would improve PNVS resolution before expanding FOV. The opinion on ANVIS was about evenly split between improving resolution and FOV. However, two cautions were emphasized by the respondees. First, these numbers do not reflect a lack of interest in increased FOV if it accompanies improved image quality. Second, the user will not accept a smaller FOV than currently provided.

The 1990 survey involved 52 ANVIS aviators from three units flying a variety of missions. Twenty of the ANVIS aviators regularly used a thermal imager viewed on a panel display in addition to ANVIS. Twenty-one PNVS aviators were also surveyed; eighteen of the PNVS aviators also used ANVIS. Again, when given an open choice of sensor, the overwhelming majority chose a pilotage system with both thermal and image-intensified imagery.

The aviators were asked to give an overall rating for PNVS and ANVIS as to adequacy of FOV and resolution (image quality); they were to answer based on their total flight experience. Table 7.2 below summarizes their answers. Seventeen of the twenty-one Apache aviators would improve PNVS resolution rather than expanding FOV with the current resolution. Fifty of the ANVIS aviators would expand ANVIS FOV if the current ANVIS resolution could be maintained.

Table 7.2.

Sensor/Feature	Good	Adequate	Inadequate
ANVIS FOV	16	45	8
ANVIS Resolution	32	36	1
PNVS FOV	2	18	1
PNVS Resolution	0	9	10

The 1992 survey was conducted after Desert Storm. No area is as devoid of distinguishable terrain features on such a scale as Saudi Arabia. The sand dunes lacked almost any vegetation and had rises and falls varying as much as 75 feet. The lack of features made the terrain relief imperceptible through the night vision sensors. This was a difficult area in which to use night vision imagers.

Of 66 aviators surveyed, 70% judged ANVIS performance in Saudi Arabia to be good or adequate. What should be noted is that the 30% inadequate rating was never experienced elsewhere. Of the 34 Apache aviators surveyed, 70% rated the PNVS performance in Saudi Arabia as good or adequate. Thermal conditions were better at the beginning of the war, and image intensifier conditions were better at the end of the war. Aviators with a choice used both systems about half the time.

The FOV of both systems was rated as adequate. Of the 34 Apache aviators, 55% rated the PNVS resolution as inadequate and 75% felt that improving resolution took precedence for a design improvement. Although image quality was a problem in Saudi Arabia, 60% of the 66 ANVIS aviators felt that improving FOV should take precedence based on their total flight experience; another 15% felt that improving FOV and resolution should take equal priority.

Flight Evaluations

In 1975, the Army conducted a flight evaluation using 2nd Generation image intensifiers. Under the illumination conditions during flight, one goggle had a 40 degree field of view with 0.6 cycles per milliradian resolution, the other had a 60 degree field of view and 0.4 cycles per milliradian resolution. The pilots rated the 40 degree goggles as more suitable for terrain flight, and the 40 degree goggles were associated with smother and more gradual control stick movements.

In 1985, a second flight evaluation was conducted. This time, the helicopter windows were masked to allow fields of view between 10° by 14° and a multiple window case with 9,000 square degrees of open window. Ground macro texture was controlled by changing locations and placing objects like tires on the ground. Micro texture was controlled by fitting the aviators with liquid crystal (LC) goggles.

Subject pilot ratings indicated that slow flight and hover can be performed with reasonable workload with a 23 by 38 degree field of view and normal visual acuity. However, adding field of view when acuity was degraded with the goggles did not improve pilot ratings. Unfortunately, the extent of the visual acuity change with the goggles was not provided, but we assume that the LC goggles provided vision of macro texture.

An experiment conducted in 1996 evaluated the impact of field of view on precision flight maneuvers. Subjects flew with FOV restricted to 40° vertical and 20, 40, 60, 80, and 100° horizontal. Normal eyesight acuity was not degraded. Maneuvers included pirouette, hovering turn, bob-up and down, precision landing, acceleration and deceleration, and slalom. Performance measures included accurate aircraft position and heading, head movement, pilot rating of flight handling qualities, and pilot rating of visual cues.

Most of the measured data showed a general increase in performance with larger FOV. Flight data indicated that performance improves with FOV up to a plateau between 40 and 80° depending on the flight maneuver. Subjective ratings of flight handling and visual cues increased with FOV up to a limit of 60 to 80° depending on task. On the basis of all the collected data, it was the researcher's opinion that the greatest overall performance gain occurred prior to the 60 to 80° FOV range under the conditions tested.

The effects of FOV and limiting resolution on flight handling were explored in two flight experiments performed in the late 1980s. Direct-view goggles were built to provide various combinations of FOV and resolution. These goggles are similar to ANVIS except they do not incorporate an image intensifier and are used during the day only. The subject pilots flew front seat cobra so the aircraft did not obstruct their view. See Figure 7.1.

Figure 7.1 shows direct view goggles fabricated to test various combinations of FOV and image resolution. The aviator subjects flew from the front seat of a Cobra.

Pilots using these goggles were asked to fly preplanned NOE and contour flight profiles. Hover and lateral flight tasks were also evaluated. In both tests, trial runs were flown without goggles to establish baseline performance levels.

Six subject pilots participated, each flying three trials of each task. Measured data included altitude, airspeed, and head motion. After each trial of each task, pilots answered questions on workload, confidence, and aircraft handling qualities. Table 7.3 shows the combinations of resolution and FOV flown on a test range at Fort Rucker, Alabama.

The term "ocular overlap" in Table 7.3 is described as follows. With 100% overlap, both eyes see the whole field of view. One technique to enlarge the display FOV while maintaining high resolution is to partially overlap the two oculars of a binocular display. With partial overlap, both eyes see the central portion of the FOV, but only one eye sees each edge of the FOV. For example, 50% overlap of a 60° goggle means that both eyes see the central 30° of the field of view. The right eye sees the right 15° of the total field of view, and the left eye sees the left 15° of the total field of view. This technique lets the optical designer reduce weight and volume by covering a large total FOV with smaller individual oculars.

Table7.3. FOV and Resolution Combinations at Fort Rucker

FOV in Degrees	Limiting Resolution	Ocular Overlap
Unrestricted	Normal eyesight	Normal
40	Normal eyesight	100 %
40	0.9 cy/mrad	100%
40	0.6	100%
40 x 60	0.9	50%
60	0.6	100%
60	0.5	100%
60 x 75	0.6	75%

The test device with 40° FOV and with 0.6 cy/mrad resolution represents current thermal imager capabilities under very favorable thermal contrast conditions. This combination also represents the capabilities of ANVIS night vision goggles under quarter moon illumination. With the exception of the device with 40° FOV and normal eyesight resolution, the other combinations shown in Table 7.3 represent achievable performance in the 1990s time frame under good thermal contrast or high light level conditions.

When FOV was held constant at 40°, decreasing resolution resulted in a substantial increase in altitude, a slight decrease in airspeed, and significantly poorer pilot ratings. Decreasing FOV to 40° but retaining undegraded visual acuity had a very minor impact on altitude and airspeed. Pilot ratings for this combination were slightly below the unrestricted baseline but were better than all other combinations tested.

With the 40° FOV, 0.6 cy/mrad device as a baseline, increasing either FOV or resolution with fully overlapped oculars improved performance and significantly elevated pilot ratings. When comparing the 40° FOV with 0.9 cy/mrad goggles to the 60° FOV with 0.6 cy/mrad device, pilots had some preference for the wider FOV but exhibited no change in performance.

Increasing FOV by diverging ocular lines of sight (that is using less than 100% overlap of the images presented to the two eyes) did not improve performance when the 40° oculars were used and caused poorer performance with the 60° oculars. The 50% partial overlap of the 40° oculars resulted in increased head motion and fatigue. Distortion for the 40° oculars was less than 1%. However, distortion in the 60° oculars reached 6%; high distortion will undoubtedly cause image convergence problems between the two eyes and lead to degraded performance.

The FOV/resolution combinations tested at Fort Rucker represented performance projected to be attainable under favorable thermal contrast or high light level conditions. A second test was flown at Fort A.P. Hill, Virginia, to explore the resolution versus field of view trade-off when simulating less than ideal thermal contrast or light level conditions.

The FOV/resolution combinations which simulated less than ideal conditions were chosen to make the flight tasks difficult but possible. The potential benefit of trading lower resolution at the edge of a sensor field of view for higher resolution at the center was also evaluated. Table 7.4 gives the combinations evaluated in the second test. Four subject pilots participated; each subject flew four trials of each task.

Table 7.4. FOV and Resolution Combinations

FOV in Degrees	Limiting Resolution
40	0.9 cy/mrad
40	0.4
40	0.5 at edge/1.1 at center
60	0.6
60	0.3
60	0.2 at edge/0.9 at center

During this test, goggle configuration did not affect altitude and airspeed performance. Once the task was defined in the baseline flight, execution did not vary significantly in terms of the airspeed or altitude which was maintained. The highest workload and lowest confidence ratings were given to the 60°, 0.3 cy/mrad goggle simulators. In this test, the pilots consistently selected the higher resolution and smaller field of view devices over the larger field of view but lower resolution devices.

If resolution at the edge of a 60º device was substantially poorer than resolution at the center, two of the pilots consistently rated the 40º field of view goggles higher, even when the 60º goggles had equivalent or better resolution in the central portion of the field of view. The other pilots rated these 40º and 60º devices as equal.

After test completion, the pilots were asked to explain this preference. The response was that, with the 60º goggles, they would see an object "and then lose it." This characteristic of the goggles was particularly bothersome during the 360º hover turn out of ground effect but also affected performance during lateral flight, NOE, and contour flight. It is likely that ocular tracking is important in the performance of all these tasks and that poor resolution at the edge of the field of view would therefore lead to adverse pilot reaction. However, ocular tracking was not measured during the test.

Summary Comments on Chapter 7

Given the improved sensitivity of the 2nd Generation system, it seems likely that aviators today would answer the thermal surveys the same way they answered the ANVIS surveys. That is, there would be an equal split between improving field of view but without degrading resolution, or improve resolution but without making the field of view smaller.

A consistent judgment of almost all Apache aviators is that both thermal and image intensified imagery are needed. Quite often, one spectral band will provide good imagery when the other is marginal, and in any environment, the two types of imager provide independent views of the flight path and surroundings.

The following observations summarize the flight test results.

- When FOV is held constant at 40 degrees, decreasing resolution resulted in a substantial increase in altitude, a slight decrease in airspeed, and significantly lower pilot ratings.
- A 40 degree FOV and undegraded visual acuity had a minor impact of altitude and airspeed when compared to

the unrestricted baseline. Pilot ratings for this configuration were slightly below the unrestricted baseline but were better than for any other configuration tested.

- When comparing 40 degree and 0.9 cycles per milliradian goggles to 60 degree and 0.6 cycles per milliradian goggles, aviators had a slight preference for the wider FOV but no change in flight performance.
- Using the 40 degree and 0.6 cycles per milliradian goggles as a baseline, improving either FOV or resolution significantly improved pilot ratings and flight performance.

Chapter 8
Partially Overlapped Oculars

Two qualities desired in a future navigation imager are wide field of view and good spatial resolution. One way to enlarge the display field of view while maintaining resolution is to partially overlap the two oculars in a binocular display. Partial binocular overlap is illustrated in Figure 8.1.

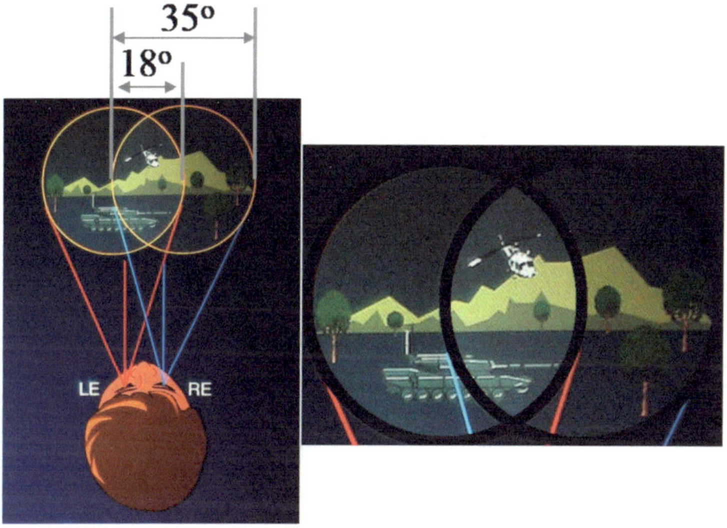

Figure 8.1. Widen FOV by having the two eyes look at different things.

In the picture at left in the figure, the left and right hand eyes both see a 35° FOV. The axes of the two oculars diverge, so that binocular vision only occurs in the 18° central portion of the field of view. Both eyes see a 35° field of view, but only 18° of that is the same region of the scene. The result is that the total field of view is 52 degrees horizontally.

The reader can confirm the following explanation by looking at a light in the ceiling and using an extended arm to block the light with his thumb. The highest contrast features in each of the variable and constantly changing binocular regions of the eye will always be present in vision. The picture at right in Figure 8.1 shows black bands around the binocular central portion of the field of view.

The left eye sees a transition from light to dark at the right hand edge of the central region. The right hand eye sees a bright to dark transition at the left hand edge of the central field of view. Our brain presents the bright central field of view with dark bands around it because each eye transitions from bright to dark at the edge of the central region, and there really are dark bands around the binocular region of the total field of view. There is nothing unknown or unusual about looning, The dark bands only extend for degree or so, and the brain then presents the light reaching each independent eye.

It is also known that binocular vision provides better contrast rendition (meaning the Contrast Thresholds are lower) than monocular vision. Most of us see the central 18 degree (^{o}) overlapped region as somewhat brighter and clearer. Also, the looning region obscures visual details from the ocular providing scene detail.

The flight experiments described in the last chapter provided evidence that partial overlapping the oculars was a bad idea. We know from open literature that partially overlapped oculars degrades target identification, and also the display concept did not perform well when tried in flight simulators. Psychophysicists have evaluated the partial overlap concept and concluded that partial overlaps of less than 40 degrees are poorly adapted to the binocular fusion characteristics of the human visual system.

However, this idea has been tried many times and failed many times, but the idea never fails to attract new advocates. If it worked, partially overlapping display oculars would provide a solution to a vexing problem. A significant widening of the helmet display field of view presents a formidable challenge, and understanding why the partial overlap configuration continually receives poor user evaluations might lead to design improvements of the partial overlap concept.

A flight experiment was designed using both an eye tracker and a head tracker to compare eye and head movement using fully overlapped and partially overlapped oculars. The overlapped and full field of views are shown in Figure 8.1. Masks were fabricated by an experienced psychophysicist based on each subject's interpupillary distance. The masks were mounted on a beam splitter attached to the aviator's helmet. Eye relief was 50.8 millimeter. See Figure 8.2. A spectral filter was applied to the beam splitter so that the eye tracking sensor did not saturate during daytime flights.

Figure 8.2 shows the eye tracker and attached field of view mask

The five subject aviators had a minimum of 3,300 and a maximum of 10,000 flight hours on helicopters, and all were ANVIS proficient. All were well trained in the variety of flight tasks executed for this experiment. Those tasks included ground effect hover, lateral flight, NOE, and contour flight.

After reducing the flight data, the ground effect hover and lateral flight tasks showed extended periods of fixations by every subject. That eye and head movement behavior was very different than the data for NOE and contour flight, and combining that data would not be representative of any flight task. Staring at a fence post for extended periods during hover or lateral flight does not provide useful data when evaluating a wide field of view display. Only the NOE and contour flight data are described here.

The other significant factor noted after initially analyzing the flight data was that the pilot labeled K kept his eyes in the center of his head. The eye and head behavior of the other four aviators were more consistent, but K's data deviated markedly. Therefore, the eye tracking data is presented two different ways. First, the K data are separated from the others, then an average of the data for all aviators are used.

The graphs in Figure 8.3 show scatter plot data amplitude versus degrees from center. Each curve is normalized to its peak, and only the data at two degrees and greater are plotted. The data labeled K is for that one aviator, and the data labeled Others combines the scatter plot data for the other four aviators. Remember that the curves are normalized and represent a fraction of peak. In other words, the lower curves have more area near zero, but showing that data changes the ordinate scale, and the off-axis data is not as visible.

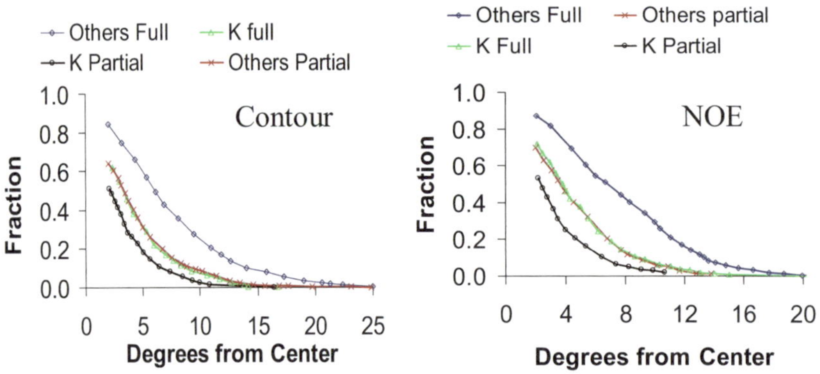

Figure 8.3 showing scatter plot data on eye degrees from center. Each curve is normalized to its individual zero degree value. Left is contour flight and right is NOE.

The plots in Figure 8.3 support two conclusions. The first is that the eye behavior of aviator K is different, much more constrained in field of view, than the other aviators. His full overlap data virtually matches the partial overlap data of the others. The second conclusion is that partially overlapping the oculars had the same effect on K as the others. On these graphs, looning occurs at 9 degrees from center, and it is clear that the eye spends much more time near center field of view when using the partially overlapped oculars, and for some reason, that is as true for K as for the others.

We know from previously published experimental data that the eye normally scans a 40 degree central field of view without head movement, and 95% of the time, no head movement will be associated with eye excursions of less than 20 degrees. The half-width of the central overlapped region tested here is only 9 degrees, and it follows that restricted eye movement should result in more head movement.

A head tracker problem resulted in getting data on only four of the subjects. For those four subjects, the head was essentially stationary 14% of the time when using the fully overlapped oculars. Their heads were stationary only 7% of the time when using partially overlapped oculars. The problem with head tracking is that we move our heads for a lot of different reasons. However, the head track data is at least consistent the contention that the head moves more when the binocular field is restricted.

Based on the average data for all subjects, the partial overlap does constrain the eye at the center of the FOV and significantly reduces the amount of time that the eye uses the outer portion of the total FOV. Averaged across all pilots and tasks, the percentage of eye fixations that occur outside the central 18° when using partial overlap was reduced by 60% ($p = 0.0170$) as compared to the full overlap (full = 24%, partial = 9%). There is no difference between tasks ($p = 0.2836$).

Looking at horizontal eye movement, the mean RMS amplitude across the five subjects for the partial overlap was only 70% of the RMS for the full overlap. This 30% reduction was significant ($p = 0.0136$). No statistically significant difference in RMS amplitude was found between tasks ($p = 0.5022$) or for the interaction between overlap and task ($p = 0.7769$). The average head velocity for partial overlap increases by 12.5% and 6% for contour and NOE flights, respectively.

The pilots indicated higher workload and lower confidence when flying the partial overlap as opposed to the full overlap. Some subjects reported nausea and fatigue after use of the partial overlap; this occurred whether the partial overlap configuration was flown first or second. There was no noticeable visual problem reported on the full overlap configuration.

Overall, these results indicate a change in characteristic head and eye motion when the partial overlap is used. There is a 10% increase in average head velocity and a significant 50% increase in the fraction of time that the head is in motion. The data suggests that the more frequent head dynamics are substituting for the lack of the ocular tracking which is restricted (60% reduction) when the partial overlap design is in use. This appears to be consistent with the hypothesis that the eyes do not track across the overlap (binocular to monocular) boundary.

The subjective data suggest that the partial overlap effectively segregates the overall 52° FOV into an 18° brighter central and two dimmer outer regions. This perceived decrease in brightness and acuity apparently derives from the lack of binocular fusion in the outer regions. The subjects indicated that looning at the overlap boundary hid scene cues; they subjectively rated the partial overlap FOV as being smaller than the fully overlapped FOV.

The flight data clearly demonstrates that the partially overlapped configuration limits the utility of the peripheral fields of view. Since no optical elements were present in the test field of view masks, the problem is fundamental to human vision. The partially overlapped FOV configuration provides a functionally smaller FOV than the fully overlapped configuration. In fact, it is not at all clear that fully overlapping the 35 degree oculars would not provide a bigger effective field of view than the partially overlapped configuration.

Chapter 9:
Video Delay, Pixel Dwell, and Field Replication

Three subjects are covered in one chapter because the flight evaluations were not experiments, only demonstrations. However, all three subjects represent critical design areas.

The Effect of Video Delays on Flight Control

Digital processing adds a delay between when the image is captured by the sensor and when it is seen by the observer. That particular delay is generally not a problem. However, there can be several digital interfaces between the imager and the pilot's display. For example, from photo array readout to image processor to avionics digital bus to symbology interface to display. If each of those interfaces is contractually granted a $1/60^{th}$ of a second frame time, the pilot's video would be a quarter second old before viewing.

Video game designers and NASA simulation engineers prefer to keep video delays at or under 40 milliseconds, but they are not dealing with the cost and complexity of developing real aircraft. That delay provides only two and a half video frames to read out the image, digital process the image, data link the image to a symbology generator, add symbology, and then display the image. One problem with accepting the necessity of that short delay requirement is that video delay is often confused with control loop delay, and we know aviators can fly systems with long control delays.

Video loop delay and control delay are very different. Older aircraft had slow control loops. The pilot moved the stick, and eventually the aircraft responded. Meanwhile, the pilot had a real time view of the aircraft frame relative to the world outside, and he learned to time his stick movements to the behavior of the airframe.

A video loop delay is different. When the aircraft moves, the pilot might sense that through his vestibular system, but he does not see the airframe to ground change until after the loop delay. With a slow control loop, the aircraft responds slowly to control input, but the visual and vestibular stimuli are concurrent, and the pilot has real time knowledge of aircraft behavior. With a video delay, the visual and vestibular stimuli occur at different times, and visual cues lag aircraft motion.

Evaluating the effects of video loop delays is best done in actual flight and not in a flight simulator. In a simulator, the major motion cues are associated with the video, not the motion base. The control and video loops become confused; there are "phantom delays" that cannot be categorized.

Two aviators participated in the flight test and alternated as subject and safety pilot. The subject pilots wore Apache helmets and viewed a helmet-mounted camera through the Apache helmet-mounted display. See Figure 9.1. The camera and display provided a 30° vertical by 40° horizontal, unity magnification image to the subject pilot. During the test, a cloth was draped over the subject's visor so that all visual cues came from the helmet display. A video digitizer provided a variable delay between camera and display. All flights were in daylight and good weather.

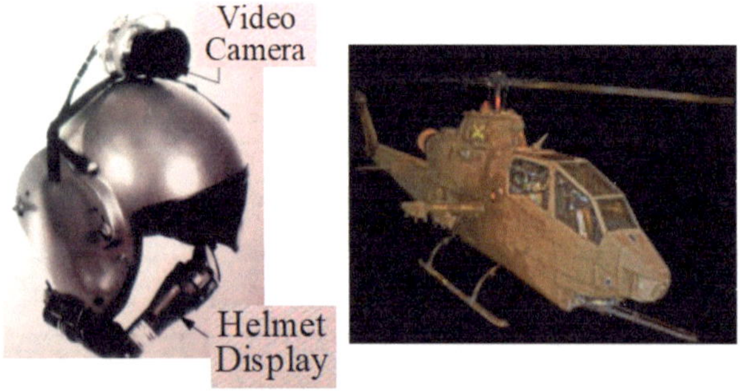

Figure 9.1 The helmet mounted camera used to capture video.

The project pilot established baselines for several, aggressive flight maneuvers using normal day, unaided vision. The maneuvers included rapid sidestep, pop-up, longitudinal acceleration and deceleration, rapid slalom, nap-of-the-earth, and contour flight. After practicing unaided and with the sensor hardware set for zero delay, the subject pilots repeated the maneuvers with the video delay increased after each iteration of the task set. Test results are based on subject and safety pilot assessments of flight performance.

On the basis of the qualitative assessment of these two pilots, there appears to be no performance impact from a 33 ms image processing delay.

Delays of 67 milliseconds were sensed by both the subject and safety pilots, but both initially stated that the delay did not cause control problems. There was some discussion after their initial statements as to whether that delay did not actually cause some slowing of the subjects control behavior, but neither subject, as pilot or as safety pilot, saw significant problems with that delay.

Delays of 100 ms or more impaired the subject pilot's ability to make stable, aggressive maneuvers. All hover tasks were more difficult; sometimes a stable hover could not be achieved. The subjects experienced the feeling that the aircraft motion was ahead of the visual scene. Alternate strategies were developed for NOE and contour to compensate for the image processing delay.

At 130 millisecond delay, the aviators had to find alternative ways to control the aircraft. They would pick a tree and fly at it, then pick another tree and fly at it. Although they were very familiar with their location, one aviator wandered off the planned course and did not realize it.

On the basis of this limited flight test, processing delays of up to 33 ms cannot be sensed by the pilot and appear to have no impact on flight performance. However, with an image processing delay of 100 ms, the pilot senses that aircraft movement is ahead of the displayed image. During these flights, and without prior training with delayed imagery, the 100 ms delay led to significant flight control problems.

These results support the advice of video gamers and NASA flight simulator operators. Video delays of 40 milliseconds or less are preferable, but delay up to 67 milliseconds (two video frames) are acceptable. Whether delays beyond about 70 milliseconds can be acceptable with training is not known, and our ability to train for long control loop delays does not answer that question.

Field Replication

The idea behind video field replication is that the imager only need operate at 30 Hertz progressive frame rate instead of 60 Hertz. The problem is that a 30 Hertz display will flicker, so each captured picture is presented twice.

However, presenting each picture twice creates a double image when there is scene to camera motion. In Figure 9.2, the camera captures 30 images per second of a moving car, but each image is displayed twice with 1/60th of a second separation. The camera operator sees a double image of the car.

Figure 9.2 When video fields are replicated, the viewer sees double.

During smooth ocular pursuit of a moving image, the observer's eye is moving with respect to the display pixels. If the pixels have long dwell times, the image will blur. If the pixels are illuminated twice, the image will appear blurred or doubled.

Full visual acuity is maintained at ocular pursuit rates of at least thirty degrees per second. If a pilot is viewing an object with relative motion of thirty degrees per second, field replication would generate a second image with a one half degree lag. That is calculated by multiplying the sixtieth of a second delay by thirty degrees per second. Since the pilot has full visual acuity, the offset would certainly be visible.

Field replication severely degrades the picture from a navigation imager used at unity magnification on a moving vehicle. However, the double images might not be all that bothersome to an engineer or manager in a laboratory or conference room.

A pilot often needs to ocular track objects at rates sufficient to cause blurring or doubling. That is, a pilot often fixates on objects moving rapidly through the sensor field of view. During terrain flight, down near the ground or trees, a pilot will often eye track a tree limb or other object in order to ascertain closure and clearance.

Also, changes in direction of gaze are made with a combination of fast saccadic eye movements and rather slower head movements. When the eye saccade is completed, the vestibulo-ocular reflex keeps the eyes fixated until head movement is complete. Since the navigation imager is head tracked, this eye-head coordination leads to frequent use of ocular tracking during scene to sensor motion.

Attempts were made to use simulations and laboratory hardware to evaluate the seriousness of the expected display artifacts. An emulation was assembled using the helmet mounted camera and display shown in Figure 9.1. The camera and display provided a forty degree, unity magnification image to the person wearing the helmet. A digitizer was used to switch between normal video and field replicated video.

This emulation clearly showed the problems associated with field replication. As stated above, however, the significance of these problems depends on the specific visual task. A flight demonstration was undertaken to gain experience in a realistic environment. The objective of this flight demonstration was to establish whether the artifacts resulting from field replication would significantly affect helicopter pilotage.

Experienced aviators flew the helmet mounted camera and concluded that field the imagery was unacceptable. Management, however, suggested that the CCD provided too good an image, making the problem worse that it would be with a thermal imager.

Fortunately, a head tracked thermal imager system that was suitable for this demonstration had recently been removed from a 6th Cavalry Blackhawk Helicopter. Those parts were borrowed back from various cooperative sources, and permission was obtained for the 6th Cav to host Apache Aviators in a front seat but with the Blackhawk flight instructor flying the aircraft. The video digitizer used with the helmet mounted camera was re-programmed, and the necessary air worthiness releases obtained. The helicopter and equipment are shown in Figure 9.3.

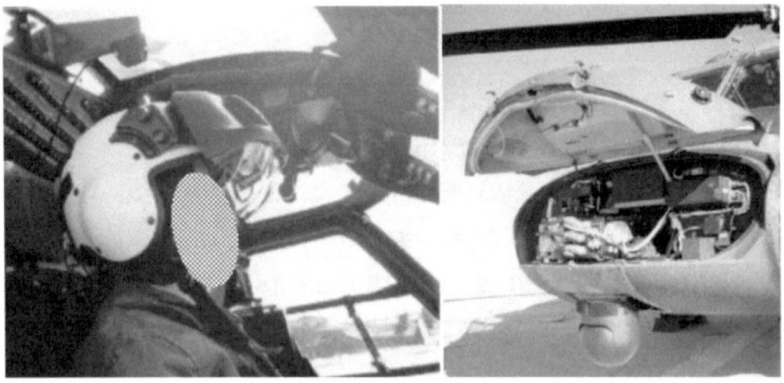

Figure 9.3 showing the thermal system used to demonstrate the effect of field replication on pilotage imagery.

The thermal imager flights with the 6th Cav at Fort Hood resulted in the same, strong, and unanimous feedback from the Apache aviators: field replication is not acceptable. Again, however, that feedback fell on deaf ears.

In the end, the fully developed imager using field replication was taken out on a balcony, and the imager panned across the parking. Observers laughed and asked "why didn't we see that?" Field replication was dropped.

Blur Due to Pixel Dwell

When a person moves his head to, say, read a roadside sign as he speeds past it, his head and eyes work together in a natural and involuntary way. Moving the head is voluntary, but it is the involuntary coordination of head and eye movements that allows the driver to read the sign.

A helmet display is attached to the helmet which is attached to the head. In one situation, perhaps the eye moves within the head to fixate a moving object, and the head moves to keep the object sufficiently centered for eye tracking. Or perhaps the person moves his head to look for a different direction of travel, then the eye saccades to find interesting locations. The eye moves within the head, and blur occurs as the eye's fixation point moves across the display pixels.

Based on published psychophysical data, humans have full visual acuity with eye movements of at least 30 and up to 60 degrees per second relative to the head. Since no loss of vision is associated with a rate of 30 degrees per second, one might assume that rate occurs when engaged in eye-head tracking of a moving object, or eye-display blur might occur when just using the head to look around.

Figure 9.1 at left shows the helmet mounted camera used in a flight evaluation of blur due to display pixel dwell; the aircraft used is shown at right in the figure. The aviators were flight qualified on the Apache PNVS, but the bi-directional scanning on the 1st Generation system made it unsuitable for this purpose. Also, the head-mounted camera eliminated turret lag. A black-out cloth was draped over the visor so that outside vision was obscured. The flight tasks and evaluation methodology are described in Chapter 7.

The graph at right in Figure 9.4 shows the modulation transfer functions (MTF) of three motion blurs. The three motion blur MTFs represent 4, 8, and 16 millisecond dwells and a 30 degree per second motion.

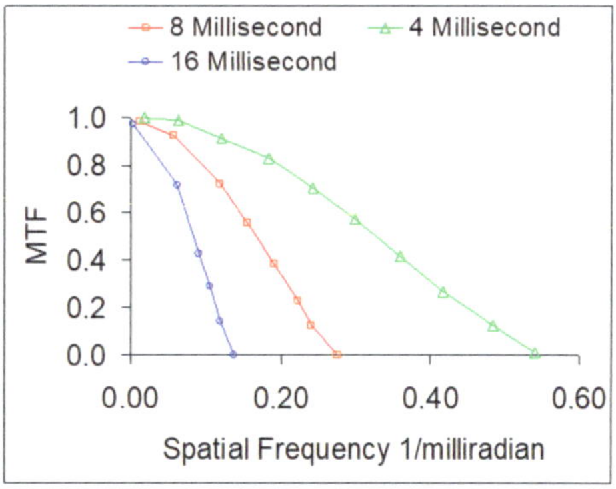

Figure 9.4 shows the MTF of motion blurs at three different dwell times.

The 4 millisecond blur did not affect flight control, workload, or meeting the standard for each flight task. The 8 millisecond blur degraded the imagery, reduced confidence, increased workload, and resulted in poor flight control as evidenced by failure to meet the flight standards. The 16 millisecond dwell degraded the imagery to the point that the camera image was deemed unacceptable for flight control.

It appears that assuming a 30 degree per second residual motion while head-tracking the scene provides a realistic assessment of the impact of display pixel dwell. Helmet mounted display pixel dwell should be limited to no more than 4 milliseconds. If a future camera improves on the AHP MTF, then a four millisecond display pixel dwell might become a limiting MTF for that future camera.

Chapter 10
Display Blur due to Vibration

If a thermal imager is available, a pilot using image intensified goggles can periodically look at the thermal image to double check his flight path. However, there is substantial vibration in a helicopter. If vibration blur prevents the pilot from seeing the details provided by the thermal imager, then the expense of that hardware will be wasted. The experiment described here addresses the concern that vibration induced display blur might nullify the performance gains expected from the new, high resolution electro-optical imagers.

Helicopter vibration produces simultaneous motion of both the eye and display; this makes a direct measurement of the display blur difficult. Further, the measurement environment is hard to specify, because vibration levels depend on aircraft type, rotor balance, and flight profile.

In order to measure display blur, a suitable aircraft was selected and the observer's contrast threshold measured under realistic cockpit and vibration conditions. The contrast loss due to display blur was estimated based on the degradation in observer contrast threshold.

The vibration environment in a particular helicopter will depend on how well the rotors are balanced and how precisely each rotor tracks the others to form a single rotor plane. Vibration level also varies with airspeed; vibration levels at hover and cruise are substantially less than at maximum speed. Ultimately, however, the worst case vibration environment is determined by the limits of pilot acceptance.

An OH-58D Kiowa Helicopter was selected for the test because it provided the closest available match to expected display location, cockpit lighting, and vibration characteristics of the helicopter in development. The test aircraft is shown in Figure 10.1; recording equipment at the right in the figure was mounted behind the subject. For flight test purposes, the rotor track and balance were set to match the worst case found in the operational OH-58D Kiowa fleet. Aviators would turn this aircraft in for maintenance.

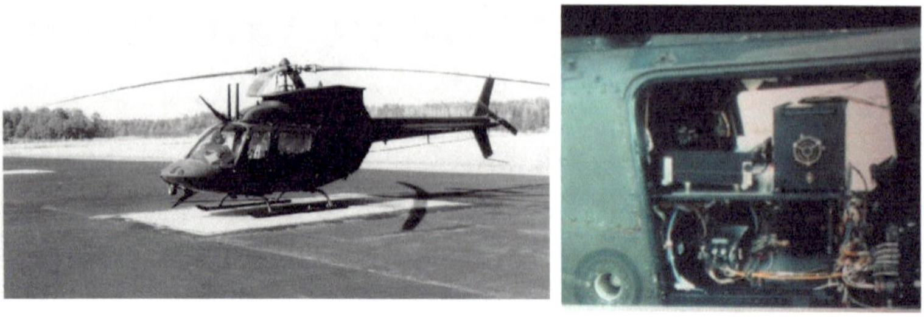

Figure 10.1 The Test Aircraft is an OH-58D with the Mast Mounted Sight removed. The accelerometer monitoring and data recording equipment was mounted in back as shown.

The objective of this flight test was to find a Modulation Transfer Function (MTF) to associate with vibration induced display blur. By selecting flight profiles of hover and 50, 70, and 120 knots level flight, data were collected under vibration conditions ranging from benign through typical worst case to extreme worst case.

Test subjects sat in the left seat and viewed the resolution charts shown in Figure 10.2. Both Air Force 3-Bar Target Resolution Charts and Rochester Institute of Technology 8 X 8 Alphanumeric Test Objects were used. Subjects recorded the smallest spatial pattern that they could resolve while the pilot flew at a designated airspeed to generate the desired vibration level. Data were collected at night in order to control ambient lighting.

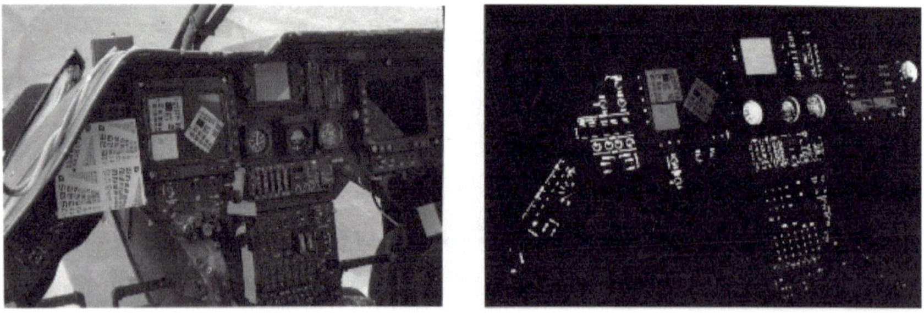

Figure 10.2 Resolution Charts Mounted on the Instrument Panel. The view at right was taken at night; the alphanumeric chart had not yet been mounted.

The bar targets were mounted on a translucent sheet and back-lit by the existing panel mounted video display. The 875 line video display uses a P-43 phosphor. The display was free-running to provide a uniform background; no raster was visible through the bar chart material.

Each 3-bar pattern consists of three equally spaced black bars; bar spacing is equal to bar width. Bar length is five times the width. Adjacent 3-bar patterns vary in frequency by about 12%. High contrast bar targets included horizontal, vertical and diagonal patterns; the photopic contrast was measured at 94% using (light - dark)/(light + dark). Low contrast bar targets included horizontal and vertical patterns; contrast was measured at 21%. Bar target peak luminance was set to 0.3 foot Lambert (fL) to conform with cockpit lighting as adjusted by the pilot.

The alphanumeric chart was used in this test primarily because it can be scored; alphanumerics provided a double check. The Alphanumeric Resolution Test Object consists of random arrays of block letters and numbers, all equally recognizable at the threshold of perception. There are four different groups of alphanumeric characters; two of the groups are arranged so that the characters are at a 90 degree angle to the other groups. The progressive change in size within the groups is about 12%.

The alphanumeric chart was taped to a metal plate which was affixed to the instrument panel as shown at the left in Figure 9.6. The alphanumeric chart was lit to 0.1 fL by the helicopter's blue-green filtered map light.

Vertical and lateral acceleration of the panel display were recorded for each flight. Vertical acceleration of the observer's seat was also recorded during each flight; the accelerometer was attached to the seat support structure. The lateral and longitudinal acceleration of the observer's seat were recorded once for each flight profile.

Five subjects participated in the flight test; two Department of the Army civilian pilots, two Army military pilots, and one engineer. Four of the five participants had 20/20 or better uncorrected visual acuity; the fifth participant had 20/20 vision with glasses. All participants had experience in testing imaging sensors. All of the pilots worked for research organizations and had extensive experience in evaluating night vision systems.

All subjects were familiar with aircraft, flight procedures, and the general test environment. All subjects had extensive experience with night vision aids and were practiced in the use of 3-bar targets to check their equipment. Prior to flight, each subject was tested using 3-bar targets at different viewing distances to insure consistency.

Subjects dark adapted for thirty minutes. Baseline data was then collected with the aircraft on ground power in a darkened hanger. The hanger was situated at the end of the airfield and was quite dark when both inside and outside lights were out. Eye to display distance was measured for each participant.

Flights occurred between 9 P.M. and midnight in calm, cool (50F) weather conditions. Flight altitudes ranged between 1,500 and 2,000 feet. Each subject performed three runs consisting of flights at hover, 50, 70, and 120 knots; speed selection was randomized within each run. The first run was used for practice and to allow the subject to fully dark adapt. The flight area was very dark, but lights could not be completely avoided while leaving the airfield.

Figures 10.3 and 10.4 show root mean square (RMS) lateral and vertical displacements, respectively, of the display and seat support. Longitudinal data was taken on the seat; worst case longitudinal displacement was less than 0.002 inches. These displacements were calculated using the average accelerations over the period that the resolution charts were being viewed; the period ranged from 50 to 80 seconds. The averages varied by about +/- 10% between runs. Averages over 2 and 8 seconds were taken to look at variations within a run; acceleration amplitude varied by approximately +/- 20%.

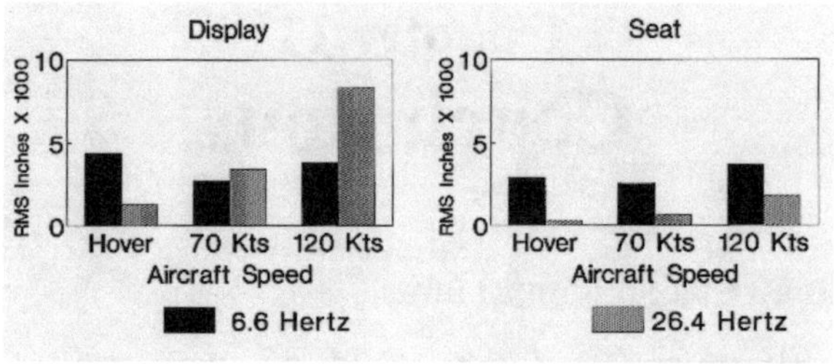

Figure 10.3 Lateral (Horizontal) Displacement of Display and Seat.

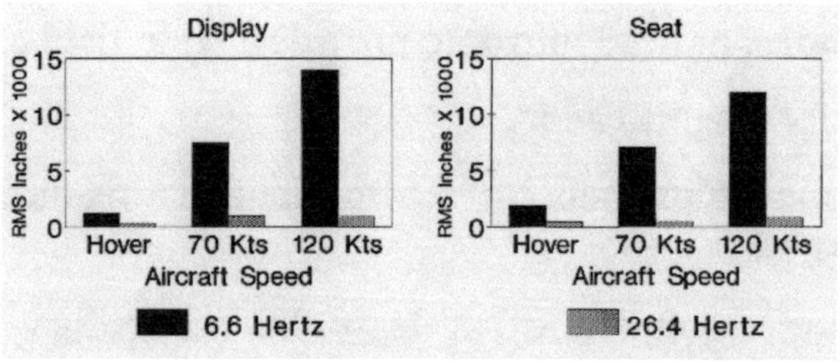

Figure 10.4 Vertical Displacement of Display and Seat.

Figure 10.5 shows MTF for lateral (horizontal) and vertical blurs. The MTF curves cover best and worst case vibrations. The MTF calculated using only the panel vibration data is not as good as the MTF in Figure 10.5, and that means that subject's eyes were, at least to some degree, moving in unison with the display.

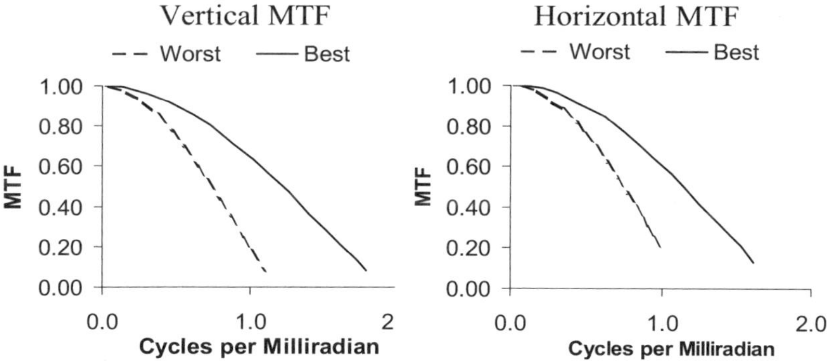

Figure 10.5 Vertical and horizontal MTF under worst and best case vibrations.

Chapter 11
Conclusions

A navigational imager should have:

- A minimum 40 degree field of view with 60 degrees desired,
- A turret or head mirror to provide a wide field of regard,
- Unity magnification,
- Motion blur cutoff > 0.6 milliradian^{-1} required and 0.9 desired,
- Static resolution of at least 0.6 milliradian^{-1} with 0.9 desired,
- Provide interpretable imagery under minimal airglow starlight in a reflective spectral band,
- Provide interpretable imagery in the long wave thermal when viewing a 0.1 K standard deviation "tree-to-ground" thermal contrast.

If the navigation imager is a stand alone system for driving or remote control of unmanned vehicles, then two additional requirements are added to the list, because ignoring either one will make the system unsuitable for most applications. The navigation imager must be affordable, and it cannot depend upon fitted, uncomfortable, eye straining, difficult to train into, head gear.

System cost becomes a concern only when the buyer wants the product. Regardless of cost, most applications cannot tolerate the requirement for an individually fitted helmet and helmet display that lets the user see the scene through one eye and requires 20 hours of intensive training to acquire beginner user skills. The helmet display in the Apache Helicopter serves many functions; it is an integral part of that weapon system. In other applications, the helmet display is an albatross.

However, a fixed forward camera viewed on a panel display does not provide the unity magnification necessary to judge distance and closure on objects in the local surroundings. The display concept in Figures 1.4 through 1.6 converts panel displays into functioning windows, so that concept can substitute for a helmet display.

Based on both user surveys and flight experiment results, the priority improvement for next generation navigation imagers would be a field of view greater than 40 degrees. However, that would not be the case if the Camera Transfer Function, including display, of the wider field of view imager fell below the AHP MTF curve in Figure 4.1. An alternative improvement that matched flight experiment results would be to maintain a 40 degree field of view but expand the Figure 4.1 abscissa scale by half. That is, an imager with a 40 degree field of view and 0.9 frequency cutoff matches the flight control capability of a 60 degree field of view with a 0.6 frequency cutoff.

None of the currently fielded navigation imagers have a motion blur problem. Image intensifiers, mechanically scanned thermal imagers, and CRT displays all have short dwell times (short exposure times). Ignoring motion blur in future systems would be a serious mistake.

Although it is well known that the long wave thermal spectral band is optimal for terrestrial viewing, the capacitor storage problem described in Chapter 5 affects long wave imagers much more that mid wave imagers. The inability to store all the photo signal generated in the long wave has led to the mistaken belief that the two spectral bands are equivalent. The capacitor problem makes the two imager types equal in sensitivity, but long wave achieves that sensitivity more than ten times faster than mid wave. Right now, with the existing capacitor problem, long wave achieves the same sensitivity as mid wave ten times faster and therefore with one tenth the motion blur.

Once the capacitor problem is fixed, the long wave imager will achieve more than three times better sensitivity than mid wave. Long wave achieves the same sensitivity as mid wave ten times faster, and given the ability to integrate longer, long wave will achieve a more than three times signal to noise advantage over mid wave.

Whether the problem is minimizing motion blur under good contrast conditions by minimizing dwell time or maximizing sensitivity under poor contrast conditions, the long wave spectral band has a significant advantage over mid wave. There are no circumstances where the mid wave and long wave spectral bands are actually equivalent for navigation imager applications.

For thermal imaging, the mechanically scanned, linear photo detector array that is currently fielded has two other advantages. One is that picture uniformity, both horizontal and vertical, is corrected real time. The second is that the system takes two samples per dwell, and the parallax motion cues so necessary for judging closure on obstacles are not aliased.

The expectation that modern, staring imager technology can improve upon fielded systems is certainly true. However, the quality of existing systems has set a high standard for a replacement, and staring imager characteristics are not as suited to the navigation application as the older, mechanically scanned design.

Part 2

Adding Color and Thermal Imagery to an Image Intensified Goggle

Under starlight illumination, there is not enough visible light to capture high resolution color video. Also, when terrain thermal contrast is weak, high resolution thermal video requires expensive cryogenically cooled imagers. Both kinds of imagery can be captured at low resolution, because large photo detectors improve signal to noise. An image intensified goggle provides a high resolution, near infrared picture of the scene. If there were a way to combine the high resolution structure of the scene with the low resolution thermal or color image, then a high resolution thermal or color picture would result.

It turns out that our eyes combine a high resolution, achromatic feature set from the retina with a low resolution color map to create the high resolution color that we perceive. The first section of Part 2 does not explain how color vision works, because no one knows how color vision works. The first section describes some known vision science and illustrates how low resolution color and thermal can be added to goggle imagery.

Technical Background

This section starts with a topic that appears to be unrelated to navigation imagers. The topic is relevant, however, because it explains how an inexpensive, uncooled, thermal imager with a long time constant can enhance image intensifier imagery.

Color Vision

In the pictures below, the fire is not visible without color, and color also tells us that the material on the ground is leaves and twigs. Color adds a visual dimension beyond gray scale.

Figure 1. Color adds a visual dimension beyond gray scale.

It is the nature of engineering that we build products for use in the near future, and more often than not, engineering decisions are not based on scientific fact, but rather best available data. Unfortunately, when it comes to human vision, vision scientists generally are unwilling to draw conclusions from the best available data. In terms of their professional careers, they are probably right in demurring.

Web Vision is a network site created to help vision scientists understand the status of work by other vision scientists. For example, an expert on the retina can catch up, in a general way, with recent science on the visual cortex. The articles are not written for a general audience, but neither do they assume that the reader is an expert in every aspect of human vision. It is not easy reading, but the site articles provide more than a collection of experimental facts.

A few years back, one of the scientists supporting that web site broke with tradition and summarized his conclusions based on available science. His article is no longer available on that site, but his conclusions seem obvious given what is known about vision. That does not mean that we, any of us, understand vision. It does mean that we can act upon the best available data rather than a popular color theory spread by word of mouth.

People see color because there are three different types of cones in the retina. The University of Rochester reflected laser light off the retinas of ten people with normal color vision to determine cone distribution; see the pictures in Figure 2 below. The pictures rendered each type of cone in a different color, the S cones are blue, the M cones are green, and the L cones are red. The pictures of the retinas of ten different people did not discover something new for vision science, but they do help the rest of us understand cone layout.

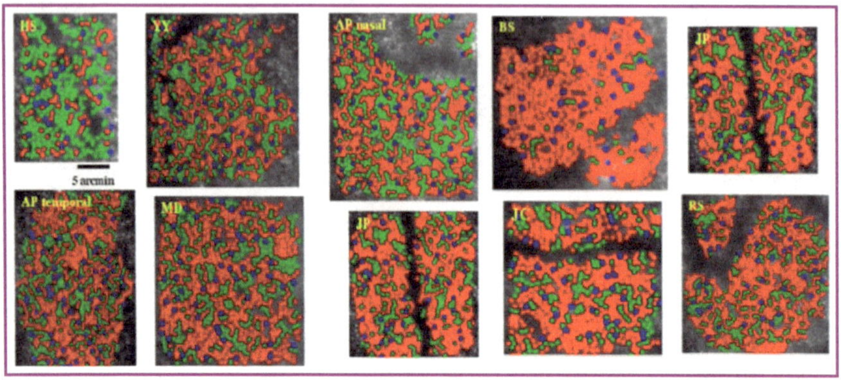

Figure 2. Colorized pictures of cones on the retina of ten people with normal color vision. The S cones are colored blue, the M cones green, and the L cones red. The gray areas are regions where the laser signals could not differentiate cone type.

The first and most important point for this discussion is that cones are not grouped in triads like color cameras. There are red, green, and blue cones, but unlike in a color camera, they do not form color cells, one type of cone in each cell. In all cases, the way the cones are distributed, it is not possible for the person's eye to directly sense high resolution color of a scene. As different as the retinas are, it is quite apparent that none of the individuals with normal color eyesight can sense full acuity red, green, and blue images. The distribution of photo detectors in the eye, the cones, does not permit it.

Further, our visual acuity necessarily means that at least the L and M cones are used together to generate a high resolution, achromatic feature set in the visual cortex. That is not just a logical conclusion; experimental evidence supports the fact that both M and L cones stimulate an achromatic feature set in the visual cortex. Perhaps the S cones also contribute, but that is an open question.

Look at the cone arrangement in each of the ten people, and imagine their eyesight if the M and L cones are not used together to create vision. The high resolution achromatic feature set has been mapped, and so has the low resolution color map. The fact that at least the M and L cones contribute to the high resolution feature set is not disputed in the vision science world. While it is true that vision scientists only understand the function of half the cell types in the visual cortex, it is also true that the entire visual cortex has been mapped by function, and there are no red, green, and blue sets of visual features. There is a high resolution, achromatic feature set formed by the combined M and L cones and a low resolution color map.

The sparcity of blue cones on the retina and the random distribution coupled with the varying ratio of M and L cones from individual to individual has been known to vision scientists for many decades. The belief that our eyes sense color using triads of cones, one blue, one green, and one red, with that triad grouping spread around the retina, is clearly wrong.

True, mapping one section of the brain does not demonstrate that some sort of encoding of blue, green, and red images is not present in other parts of the brain. But, given the layout of the retina, how might the brain create three feature sets? In addition, Edwin Land demonstrated a half century ago that color is not sensed point by point on the retina; he demonstrated that color at each point in a scene depends on the whole scene.

Two things are true. First, no one knows how the brain works. Second, the popular, almost universally accepted, theory that the eyes capture red, green, and blue images in much the same manner as color television, is simply not true. Our brains take a high resolution, achromatic feature set and combine it with a low resolution color map to create a high resolution color image. As engineers, we can take advantage of the properties of color vision.

In Figure 3, the woman is viewing a high resolution, achromatic picture through a low resolution, transparent color image. She sees a high resolution, color image of the parrot. Of course, we can also simply multiply images A and B pixel intensities and display the result on C.

Figure 3. A high resolution monochromatic picture (A) is viewed through a semi-transparent, blurred color picture (B). The person sees a high resolution color image (C). Multiplying A and B pixel by pixel gives the same result.

Figure 4 shows a picture of a pool area selected because the pool has tile with straight lines and the scene has a variety of colors. Picture A shows two transparencies held up against a blank, white computer screen. One transparency, photographed in B, has a high resolution, monochromatic image of the pool area. The second transparency, photographed in (C), has a blurred color image of the area. Photograph D shows the two transparencies held together against the white screen. The color picture is high resolution when the two transparencies are viewed together. Pictures E and F show blow ups of D and C to illustrate that the tile lines are now straight, and the picture has good color resolution.

In Figure 5, Item 1 is an LCD display with the backlight removed; it has a low resolution color picture of a Mandrill on it. Item 2 is a laptop. Picture 3 was taken with a color camera when the color Mandrill was backlit with a blank white area. Picture 4 was taken with a high resolution, black and white picture of the Mandrill on the laptop. This figure illustrates the hardware configuration that will be used in the navigation imager. When the high resolution photo of the Mandrill backlights the low resolution color, a high resolution color picture appears.

Figure 4 illustrates that the eye creates a high resolution color image when a low resolution color picture is viewed through a high resolution achromatic picture.

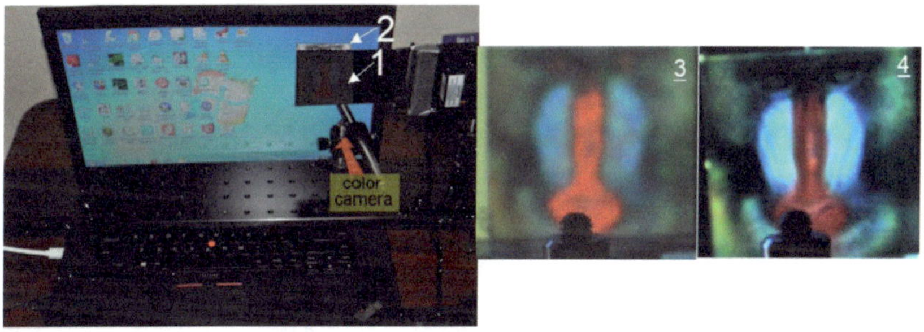

Figure 5 illustrates that putting a low resolution color image on an LCD and backlighting it with a high resolution, monochromatic image creates a full color, high resolution picture.

Color and thermal are alike in the sense that worst case conditions result in very weak ground radiance. Ground illumination in the visible is an order of magnitude down form the near infrared where image intensifiers operate. Also, the visible spectrum is divided into three color bands, further reducing available signal. If we try to operate a high resolution color camera under starlight conditions, the photo detectors will be small, the noise bandwidth high (because there are a lot of pixels), and the signal to noise very poor.

On the other hand, if we design a color camera with very large pixels, then each pixel integrates much more signal, and the electrical noise bandwidth narrows. Further, and this point is critical for navigation imagers, we can integrate signal for a full frame time without worrying about motion blur. The color pixels are huge and dwarf the size of motion blur.

Consider the implications. The color image is formed by combining the sensitivity of very large, long dwell time pixels with the resolution of an image formed by sensing near infrared radiation. It would not be possible to capture the resulting color image by sensing only in the visible spectral band.

Our color vision mechanisms provide a means to add spatial resolution to low resolution color and thermal pictures. The image intensifier is a reliable and inexpensive means of capturing high resolution night imagery, although it is well known that monochromatic reflective imagery can often be bland, and that both color and thermal enhance our ability to discriminate scene details.

It is also true that photon counting thermal imagers require long dwell times to achieve good sensitivity, and that uncooled thermal tends to have long time constants. Also, the best value and easiest to use thermal imagers are uncooled, but it is not clear that they meet the minimum required sensitivity requirement if the detectors are small.

Combining a low resolution uncooled thermal imager with the image intensifier solves multiple problems. The bland nature of reflective imagery in some environments is enhanced. The sensitivity and dwell time problems with the inexpensive thermal imager are solved, because large pixels provide good signal to noise, and motion blur has little effect on a low resolution image.

Figure 6 shows an image intensified picture at the top that is processed like Figure 3 with a 128 by 94 pixel, uncooled thermal image. The processed image is shown at the bottom of the figure. The thermal image enhances the scene without hurting picture sharpness.

Figure 6 shows color processing using a low resolution (128 by 94 pixel) thermal image as the color red.

PVS-14 with Color and Thermal Imaging

Currently fielded and available night vision goggles operate in the near infrared (NIR) spectral band from about 0.6 to 0.9 micrometer wavelength. At night, NIR illumination is greater than in the visible spectrum, and the NIR provides an achromatic image with good spatial resolution and, generally, a high signal-to-noise. Further, commercial PVS-14 monocular intensifiers now have a white display phosphor.

Figure 7. shows a PVS-14 with a color camera and LCD added. The picture shows the line-of-sight of the color camera and image intensifier as side by side, but an optical splitter mounted n the front of the PVS-14 might be necessary.

The components in a typical 3rd Generation PVS-14 are illustrated at the top of Figure 8. Light imaged on the cathode generates photo electrons that are attracted to the microchannel plate (MCP) where electrons are multiplied. After exiting the MCP, the electrons are accelerated to the phosphor where an image is formed. A fiber optic twist erects the image for view via an eyepiece.

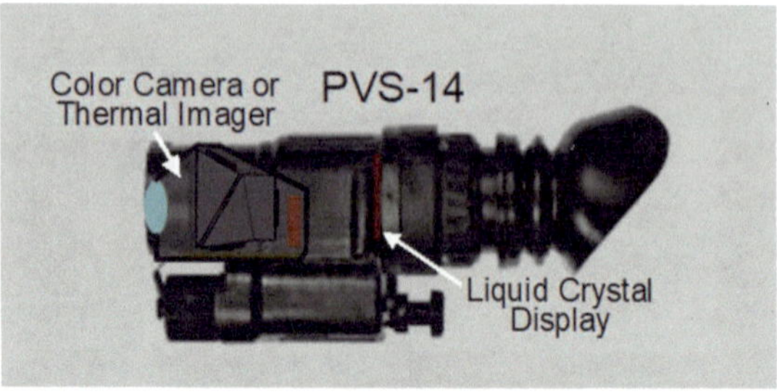

Figure 7. A PVS-14 with color camera added.

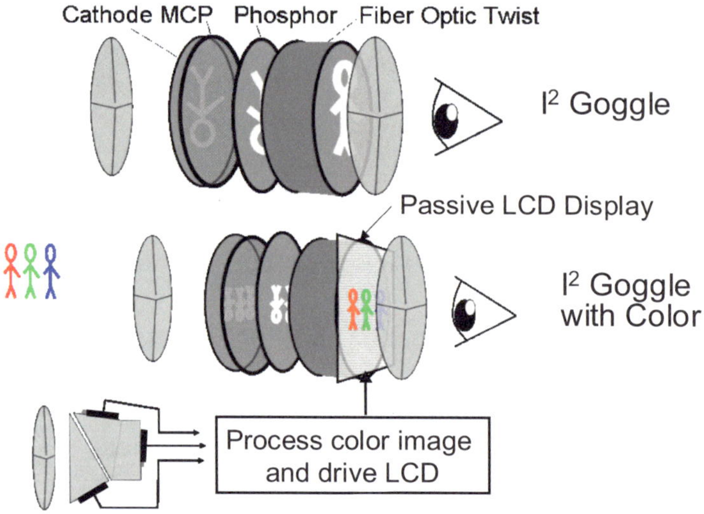

Figure 8 shows PVS-14 components at the top and the components needed to add color at the bottom of the figure.

An LCD can be added at the end of the fiber optic twist. A passive LCD is needed. Although the eyepiece is focused on the intensifier image, any structure visible in the LCD will distract the eye. Also, there is no reason to expect that the speed or dynamic range of an active LCD is needed.

The LCD display has 512 by 512 pixels in order to ensure color uniformity. The three silicon focal plane arrays are shown on a Philips Prism, each array with 128 by 128 pixels that are 70 microns on edge. There is not enough light to use one, Bayer Pattern color detector array. However, although the Philips Prism provides the best performance, the quality imagery provided by that selection is likely not needed, and a lighter design might be substituted. Figure 9 shows focal plane placement and electronics.

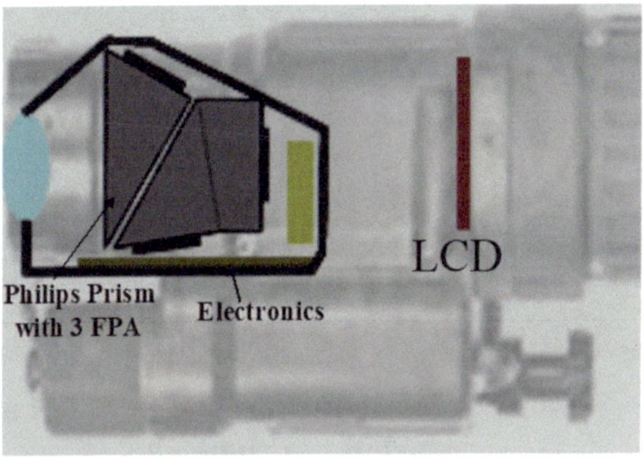

Figure 9. Focal plane placement and electronics card.

Table 1 shows camera signal for each color spectral band under worst case starlight or typical overcast starlight. Starlight with thick, thunderstorm clouds would require some artificial illumination.

Figure 10 is a simulated picture of Goldhill under the Table 1 illumination conditions. The calculations are for F/1 optics and the field of view matches the PVS-14. A full frame dwell is assumed. As regards motion blur and signal to noise, remember that these focal planes have very large photo detectors, and the picture is provided by the image intensifier.

Table 1. Signals under worst case starlight.

Band (μ)	Photons / detector	Average Photons	Signal (0.2 contrast)	Pixel signal to noise
0.4 -0.5	240	120	24	2.1
0.5 - 0.6	460	230	46	2.9
0.6- 0.7	520	260	52	3.1

The integration described here is for a color camera, but a low resolution, uncooled, thermal imager with large pixels for sensitivity would substitute for a single color. Figure 6 is an actual picture combining PVS-14 and uncooled thermal images, although streetlights down the road from the farm provided substantially more than starlight illumination for the intensifier.

Figure 10 showing Goldhill during the day at left and under starlight at right.

All of the parts can be bought on the commercial market, although the focal plane and LCD are not off the shelf. All of the technologies needed have been around for decades, and the development is low risk from a hardware fabrication standpoint.

In terms of technical risk, the calculated color signals are certainly not strong, but airglow has been studied rather thoroughly because it interferes with astronomy, and the calculation is for worst case. Also, the actual pictures in Figures 4, 5, and 6 demonstrate the viability of the method.

Perhaps the biggest question is whether there is a market for adding color and or thermal to a monochromatic picture. Our ability to see color suggests that there is a natural advantage to colored imagery. Also, the military has learned that thermal provides visual cues not available in monochromatic reflective imagery.

Whether for the goggles or other applications, the primary benefit of the implementation described here is that images from focal plane arrays with a small number of large photo detectors can be combined with the achromatic image captured in a different spectral band to create color and thermal imagery with the sensitivity of the large detectors but the resolution of the achromatic image. The process makes high resolution thermal imagery inexpensive, and it makes high resolution color under starlight illumination possible.

Bibliography

1. Electro-Optical Imaging: System Performance and Modeling, Edited by Lucien Biberman, Chapter 26, Sensor System Psychophysics, Richard Vollmerhausen, SPIE, 2000.
2. Wiley Encyclopedia of Electrical and Electronics Engineering, Vol. 8, Ed. John G. Webster, "Helicopter Night Pilotage," Richard H. Vollmerhausen, 1999.
3. Richard H. Vollmerhausen, "Representing the observer in electro-optical target acquisition models," Opt. Express 17, 17253-17268 (2009)
4. "Night illumination in the visible, NIR, and SWIR spectral bands," Richard H. Vollmerhausen and Tana Maurer, SPIE Vol. 5076, 2003.
5. "The Effect of Helicopter vibration on the Viewing of Panel Mounted Displays," R. Vollmerhausen and K. Mayes, SID Digest, Vol. XXVI, pp 651-654, 1995.
6. "The effect of Sensor Field Replication on Displayed Imagery," R. Vollmerhausen and T. Bui, SID digest, Vol. XXV, pp 667-670, 1995.
7. "Overlap Binocular Field-of-View Flight Experiment," T. Bui, R. Vollmerhausen, and B. Tsou, SID digest, Vol. XXV, pp 306-308, 1994.
8. "Display of Sampled Imagery," R. Vollmerhausen, IRIA-IRIS Proceedings n Passive Sensors, Vol. 1, pp 175-192, 1990.
9. "Design Criteria for Night Pilotage," R. Vollmerhausen, Passive IRIS, 1990.
10. "Design Criteria for Night Pilotage Sensors," R. Vollmerhausen and C. Nash, American Helicopter Society, 1989.
11. "Design Criteria for Night Pilotage Sensors," R. Vollmerhausen and C. Nash, Army Science Conference, 1988.
12. "Night Vision Pilotage System for Attack Helicopters," R. Vollmerhausen, C. Nash, Passive IRIS, 1986.
13. "Modeling the blur associated with vibration and motion," Richard Vollmerhausen, Mel H. Friedman, Joe Reynolds, and Stephen Burks, Proc. SPIE 6543, 6543OU, 2007.
14. "New weather depiction technology for night vision goggle (NVG) training," Scott Theleman, Jennifer Hegarty, Richard Vollmerhausen,

Courtney Scott, John Schroeder, Frank P. Colby, and S. Napier, Proc. SPIE 6303, 63030F, 2006.

15. "New weather depiction technology for night vision goggle (NVG) training: 3D virtual/augmented reality scene-weather-atmosphere-target simulation," Michelle Folaron, Martin Deacutis, Jennifer Hegarty, Richard Vollmerhausen, John Schroeder, and Frank P. Colby, Proc. SPIE 6557, 65570C, 2007.
16. "Using a targeting metric to predict the utility of an EO imager as a pilotage aid," Richard H. Vollmerhausen and Trang Bui., Proc. SPIE 6207, 62070C, 2006.
17. "Head Movements During Contour Flight," USAARL Report No. 87-1, Robert Verona, Clarence Nash, William Holt, John Crosley, October, 1986.
18. "In-Flight Simulation of Field of View Restrictions on Rotorcraft's Pilot Workload, Performance, and Visual Cueing," Loran Haworth, Zsolt Halmos, American Helicopter Society, 52 Annual Forum, 1996.

Made in the USA
Monee, IL
05 November 2026

67811037R00066